SpringerBriefs in Mathematics

SpringerBriefs present concise summaries of cutting-edge research and practical applications across a wide spectrum of fields. Featuring compact volumes of 50 to 125 pages, the series covers a range of content from professional to academic. Briefs are characterized by fast, global electronic dissemination, standard publishing contracts, standardized manuscript preparation and formatting guidelines, and expedited production schedules.

Typical topics might include:

A timely report of state-of-the art techniques A bridge between new research results, as published in journal articles, and a contextual literature review A snapshot of a hot or emerging topic An in-depth case study A presentation of core concepts that students must understand in order to make independent contributions

SpringerBriefs in Mathematics showcases expositions in all areas of mathematics and applied mathematics. Manuscripts presenting new results or a single new result in a classical field, new field, or an emerging topic, applications, or bridges between new results and already published works, are encouraged. The series is intended for mathematicians and applied mathematicians. All works are peer-reviewed to meet the highest standards of scientific literature.

Titles from this series are indexed by Scopus, Web of Science, Mathematical Reviews, and zbMATH.

Takeo Ohsawa • Thomas Pawlaschyk
Editors

Analytic Continuation and
q-Convexity

 Springer

Takeo Ohsawa
Graduate School of Mathematics
Nagoya University
Nagoya, Japan

Thomas Pawlaschyk
Scl. of Mathematics and Natural Sciences
University of Wuppertal
Wuppertal, Germany

ISSN 2191-8198 ISSN 2191-8201 (electronic)
SpringerBriefs in Mathematics
ISBN 978-981-19-1238-2 ISBN 978-981-19-1239-9 (eBook)
https://doi.org/10.1007/978-981-19-1239-9

This Springer imprint is published by the registered company Springer Nature Singapore Pte Ltd.
The registered company address is: 152 Beach Road, #21-01/04 Gateway East, Singapore 189721, Singapore

To the memory of Professor Osamu Fujita

Preface

By investigating q-cycles, the first-named author of this monograph obtained a result on the analytic structure of graphs of certain mappings whose complements are q-convex. After having presented his result at the Pohang University of Science and Technology, South Korea, in summer 2019, one of the attendants, Nikolay Shcherbina, mentioned that, using completely different techniques, a more general statement was already proved in the doctoral thesis by his former PhD student in 2015, the second-named author of this monograph. Since—as often—teaching duties slowed down further research on this topic, it has not (yet) been published in a journal. Anyway, the idea came up to present the statement along with a survey on generalized pseudoconvexity at a conference at Osaka City University, Japan, in early 2020, organized by the first-named author and Takayuki Koike. While we prepared the lectures and arranged the travel details, the world followed with concern the news on a new dangerous virus, Covid-19, which quickly spread from the world's new economic global player, China, to all over the world in a few months. A global pandemic started, and each country was forced to initiate lockdowns, such as shutting down borders to protect their own inhabitants. Everyday life and economy reduced to a minimum with the only goal to keep the system running and overcome the pandemic with patience and endurance. Eventually, travel plans were overthrown, and the organizers were forced to postpone the conference in Osaka. Following the popular saying "If one door closes, another one opens!" (however this is possible physically), research collaboration continued thanks to the wide scope of modern digital communication tools and high-speed Internet. Finally, the conference was held in the form of a virtual online meeting in February 2021, while the writing of this monograph was supported by already old-fashioned but nevertheless digital methods: email, text editor, and LaTeX compiler.

The focus of our monograph lies on the further development of the classical achievements in analysis of several complex variables, in particular, the analytic continuation and the analytic structure of sets in a setting in which generalized convexity plays a crucial role. The starting point was the question whether holomorphic functions extend analytically their domain of definition, which gave birth to the notion of a domain of holomorphy. It turned out that these domains are determined by holomorphic convexity and pseudoconvexity. The latter notion involves plurisub-harmonic functions which are defined via harmonic and subharmonic functions and resemble convex functions in the complex setting. This notion generalizes to q-pseudoconvexity in the non-smooth and q-convexity in the smooth case. In fact, both are equivalent in the smooth case up to an index shift by 1 and occur naturally in the study of complements of analytic varieties. However, because of the non-smooth character of the q-pseudoconvexity, it allows general basic research, while, because of smoothness, the q-convexity is suitable for studies up to complex spaces and in the context of cohomology groups. Since the index q indicates the number of negative eigenvalues of the complex Hessian of certain functions, q-convexity interacts with Morse theory. Although q-convexity was studied intensively during the second half of the twentieth century, it is hard to find an overview of the development of q-convexity and a collection of the most important results in the literature. Since recently, studies on generalized convexity experienced a revival in complex analysis and represent an important part of our research, we felt motivated to gather our knowledge on this topic in a collective volume. The outcome is the present monograph which might be interesting for a wider academic readership: for undergraduates it might serve as an orientation, graduates can benefit from a concise collection of classical results and references for their own studies, while experts can learn about the state of the art and recent developments on q-convexity.

Nagoya, Japan   Takeo Ohsawa
Wuppertal, Germany   Thomas Pawlaschyk
2021

Acknowledgments

The manuscript was planned to be read in the workshop *Studies on pseudoconvexity of general order*, which was canceled because of the expansion of Covid-19.

The first-named author thanks Mihnea Colțoiu and Cezar Joița for informing the author that Remark 4.1 was first due to M. Peternell [144] by sending him a copy of Barlet and Magnússon [44] in these difficult circumstances.

The second-named author thanks his former supervisors Nikolay Shcherbina and Eduardo Zeron for their support during the writing of his PhD thesis at the University of Wuppertal, where the content for the first three chapters of this monograph originates. Thanks also to Kang-Tae Kim for his hospitality during the authors' multiple research stays at Postech, Pohang, South Korea. It was the perfect atmosphere for the authors' further research on q-pseudoconvexity.

The authors wish to thank the referees for valuable comments and suggestions which improved the presentation of this monograph.

Contents

Introduction

Questions on analytic continuation have been the driving force in the development of complex analysis in several variables. Just as the curvature notion plays an important role in Riemannian geometry, the concept of pseudoconvexity is central in the study of analytic continuation. The Levi problem on pseudoconvex domains was one of the most difficult questions whose solution by Oka [134, 135], Bremermann [24], and Norguet [128] has opened a new area in complex analysis. Since then, methods of the sheaf theory and partial differential equations (PDE) have been brought into the scope. The sheaf theory has also become a basic tool in algebraic geometry, and the method of PDE yielded precise and quantitative results which have been applied to the existence questions in complex geometry.

In the current topics of several complex variables continuing these classical achievements, there has been a lot of activity on the domains with generalized pseudoconvexity, particularly on q-convex domains in the sense of Rothstein [150] and Grauert [73]. Recently, a well-known result of Hartogs [84] was generalized in terms of q-convexity in [133] and [139, 140], independently and by completely different methods.

The present monograph is arranged as follows: in Chap. 1, we introduce the basic notions and classical results in complex analysis of several variables since the beginning of the twentieth century. In Chap. 2, we discuss the q-plurisubharmonic functions in the sense of Hunt and Murray [97] and these properties which are used to define q-pseudoconvex sets in the sense of Słodkowski [160] in Chap. 3. In the same chapter, we list equivalent notions of q-pseudoconvexity. In fact, those of Rothstein and Słodkowski are indeed equivalent. Moreover, we change the perspective and study complements of q-pseudoconvex sets, the q-pseudo*concave* sets, which we use in one of the generalizations of Hartogs' theorem in Theorem 3.7 (cf. [139, 140]). Finally, in Chap. 4, we introduce the notion of q-convexity (and q-completeness) in the sense of Grauert and study its relation to cycle spaces in the complex projective space. It yields the main statement, Theorem 4.10, which implies another generalization of Hartogs' theorem in Corollary 4.2 (cf. [133]).

Chapter 1
Analytic Continuation and Classical Pseudoconvexity

Every introduction to complex analysis of several complex variables clarifies sooner or later the relation between domains of holomorphy and pseudoconvex domains, the latter defined by plurisubharmonic functions. In the first half of the twentieth century, intense research on these objects accumulated in the groundbreaking proof of Levi's conjecture [116] by Oka in 1942 on the equivalence of these two types of domains [135].

In this chapter, we roughly reconstruct the historical development and summarize concisely the very classical statements on this topic. We do not claim completeness, but only recall those results which are important to us in the forthcoming chapters in which we investigate q-pseudoconvex and q-convex domains and their application to analytic continuation.

For more details, we refer the interested reader to the rich collection of classical textbooks on complex analysis in several variables written by mathematicians from all around the world, such as Grauert and Remmert [78], Hörmander [95], Krantz [112], Narasimhan [124], Nishino [127], Shabat [153] and Vladimirov [166], to name just a few. For a rich source of examples and counterexamples which specify the relation of the forthcoming objects we recommend Fornæss' and Stensønes' book [61]. We also recommend Remmert's survey filled with anecdotes on the historical development of complex spaces [148].

1.1 Domains of Holomorphy

Holomorphic functions constitute the fundamental objects in complex analysis. Given an open set U in $\mathbb{C}^n$, a function $f : U \to \mathbb{C}$ is called *holomorphic* if it is complex differentiable at each point $p \in U$, i.e., it fulfills the partial differential

© The Author(s), under exclusive license to Springer Nature Singapore Pte Ltd. 2022
T. Ohsawa, T. Pawlaschyk, *Analytic Continuation and q-Convexity*, SpringerBriefs
in Mathematics, https://doi.org/10.1007/978-981-19-1239-9_1

equation

$$\frac{\partial f}{\partial \bar{z}_j}(p) = 0 \quad \text{for every } j = 1, \dots, n$$

or, equivalently, $\bar{\partial} f = 0$. The class of holomorphic functions on U is denoted by $O(U)$. A function f is holomorphic if and only if it is *analytic* in the sense that, locally near each $p \in U$, f has a power series expansion,

$$f(z) = \sum_{\alpha} a_{\alpha}(z - p)^{\alpha},$$

where α denotes a multi-index with n components, each ranging between all non-negative integers. The study of the radius of convergence of analytic functions immediately leads to the question whether and how far f extends analytically beyond its initial domain of definition. Now assume that the open set U lies inside some larger open set V in $\mathbb{C}^n$. If there exists a holomorphic function $F : V \to \mathbb{C}$ such that $F|_U = f$, then f *extends analytically into* V and F is an *analytic continuation* of f into V.

At the beginning of the twentieth century, Hartogs devoted a series of papers [82–84] on the analytic continuation of power series of the form $f(z, w) = \sum_{\nu} f_{\nu}(z)w^{\nu}$ in $\mathbb{C}^2$. His work merged into what is still sometimes called *Kontinuitätssatz*. A modern interpretation reads as follows. Let $\Delta = \{z \in \mathbb{C} : |z| < 1\}$ denote the unit disc in $\mathbb{C}$. Given any domain $D \subset \mathbb{C}^n$, $n \geq 2$, on which there exists a holomorphic function that cannot be continued analytically across any boundary point of D, the image of a holomorphic map $\Phi : \Delta^n \to \mathbb{C}^n$ is contained in D if already $\Phi(H)$ lies in D, where H stands for the *Hartogs figure*

$$H = \Delta^n \setminus \left\{z \in \Delta^n : |z_1| \leq \frac{1}{2} \text{ and } \max\{|z_j| : 2 \leq j \leq n\} \geq \frac{1}{2}\right\}.$$

Since its original version was not suitable for manifolds, it was reformulated by Behnke [20] and later by Behnke and Sommer [22] in the following manner by means of analytic varieties (cf. [153]).

Theorem 1.1 (Continuity Principle) *Let $\{S_{\nu}\}_{\nu}$ be a sequence of analytic surfaces such that the closure of each S_{ν} is contained in a domain D in $\mathbb{C}^n$. If $\{S_{\nu}\}_{\nu}$ converges to a set S and the boundaries $\{\partial S_{\nu}\}_{\nu}$ to a set $T \Subset D$, then any function $f \in O(D)$ extends analytically into some neighborhood of S.*

In the planar case $n = 1$, the function $h(z) = 1/z$ is holomorphic outside the origin and, obviously, has no analytic continuation into the whole of $\mathbb{C}$. However, Riemann's theorem on removeable singularities asserts that, if a function $f : \Delta \to \mathbb{C}$ is holomorphic exactly on the punctured disc $\Delta \setminus \{0\}$ and is at least bounded near the origin, it extends analytically into the whole of Δ.

The situation completely changes in higher dimensions if $n \geq 2$. By applying the Cauchy integral formula on each variable of a given function $f = f(z, w)$ that is analytic in a neighborhood of the bidisc $D = \Delta \times \Delta$ in $\mathbb{C}^2$, we have that

$$f(z, w) = \left(\frac{1}{2\pi i}\right)^2 \int_{\partial\Delta} \int_{\partial\Delta} \frac{f(\xi, \eta)}{(\xi - z)(\eta - w)} d\xi d\eta$$

for each $(z, w) \in D$. Hartogs observed in [83] that this identity already holds true if f is not analytic in a neighborhood of the whole of D, but only on some smaller open subset containing the boundary of D. His investigation led to the well-known Kugelsatz[1] which is expressed nowadays as follows.

Theorem 1.2 (Hartogs' Extension Theorem) *Let K be a compact set inside a domain D in $\mathbb{C}^n$, $n \geq 2$, such that $D \setminus K$ is connected and let $f : D \setminus K \to \mathbb{C}$ be holomorphic. Then f extends analytically into D, i.e.,*

$$O(D \setminus K) = O(D).$$

In particular, every holomorphic function on $\{z \in \mathbb{C}^n : r < |z| < R\}$ extends holomorphically into the entire ball $B = B_R^n(0)$ if $n \geq 2$. Moreover, Hartogs showed in the same paper [83] that if a function f is analytic near a two-dimensional Hartogs figure, it is already analytic in a neighborhood of the bidisc Δ^2. Eventually, the sets which do or *do not* permit analytic continuation for all of its holomorphic functions became of special interest in complex analysis.

Definition 1.1 Let D be a domain in $\mathbb{C}^n$. Then D is a *domain of holomorphy* if there are *no* non-empty open sets U and V in $\mathbb{C}^n$ with the following properties:

1. $U \subset D \cap V$;
2. V is connected and not contained in D;
3. for every $f \in O(D)$ there exists $h \in O(V)$ such that $f = h$ on U.

One of Hartogs' results [84] stands in our focus. It clarifies the relation of analytic graphs and domains of holomorphy.

Theorem 1.3 (Hartogs' Theorem on Analytic Continuation) *Let $f : D \to \mathbb{C}$ be a continuous function defined on a domain of holomorphy $D \subset \mathbb{C}^n$. Then the complement of its graph*

$$G_f = \{(z, \zeta) : z \in D, \zeta = f(z)\}$$

in $D \times \mathbb{C}$ is a domain of holomorphy if and only if the function f is holomorphic.

---

[1] "Satz" is the German expression for "Theorem", as everyone certainly knows, and "Kugel" is one of the German words for "ball". If the reader wonders how to pronounce "Kugel", here is a modern German saying on the state of our current society: "Die Welt ist eine Google."

Notice that in his original paper, Hartogs assumed the function f to be merely continuous, but this is not relevant to our purposes. We will encounter different versions and generalizations later in Theorem 1.7, Theorem 1.11 and as a special case in our main statements of this monograph, Theorem 3.7 and Corollary 4.2. For a proof of Hartogs' theorem using modern techniques we recommend for instance Shabat's book (cf. [153, Chapter III.42]).

As we have seen in Hartogs' extension theorem, domains of the form $D \setminus K$ can never be domains of holomorphy, but every domain D in $\mathbb{C}^n$ is contained inside a largest (Riemann) domain to which all holomorphic functions on D extend (cf. [124] for a sheaf theoretic and [100] for a more effective construction of the so-called *envelope of holomorphy*). However, in the complex plane, every domain $G \subset \mathbb{C}$ is a domain of holomorphy, since for every boundary point $p \in G$ the function $h_p(z) = 1/(z - p)$ constitutes a holomorphic function which does not extend analytically near p. In higher dimensions, Cartan and Thullen [35] proved in 1932 that every domain of holomorphy D in $\mathbb{C}^n$ is holomorphically convex in the sense that, for any closed discrete set $\{p_\mu; \mu \in \mathbb{N}\} \subset D$, one can find a holomorphic function f on D such that $\lim_{j \to \infty} |f(p_j)| = \infty$. Nowadays, their result is stated as follows.

Theorem 1.4 (Cartan–Thullen) *Let D be a domain in $\mathbb{C}^n$, $n \geq 2$. Then the following properties are equivalent:*

1. *D is a domain of holomorphy.*
2. *For every compact set K in D we have that*

$$\mathrm{dist}(\widehat{K}_D, \partial D) = \mathrm{dist}(K, \partial D),$$

   where $\widehat{K}_D := \{z \in D : |f(z)| \leq \max_K |f| \ \forall \ f \in \mathcal{O}(D)\}$ is the holomorphically convex hull of K in D.
3. *D is holomorphically convex, i.e., for every compact set K in D its hull $\widehat{K}_D$ is compactly contained in D.*

   Here, $\mathrm{dist}(A, B) := \inf\{|z - w| : z \in A, \ w \in B\}$ for two sets A, B in $\mathbb{C}^n$.

From this, it easily follows that a finite intersection of domains of holomorphy remains a domain of holomorphy, but for unions this is no longer true in general. Nevertheless, at least the following statement, due to Behnke and Stein [21], holds true and turned out to become an important tool to examine domains of holomorphy.

Theorem 1.5 (Behnke–Stein) *The union of a countable, increasing sequence $\{D_j\}_{j=1}^{\infty}$ of nested domains of holomorphy $D_j \subset D_{j+1}$ is again a domain of holomorphy.*

The question arises whether there exist more tools which allow us to create domains of holomorphy in a simple way. Indeed, inside the definition of the holomorphically convex hull, we encounter the function $z \mapsto |f(z)|$, where $f : D \to \mathbb{C}$ is holomorphic. It belongs to the family of the so-called *plurisubharmonic* functions, which open gates to a series of further characterizations from a very different perspective, closely related to convexity. We will focus on these objects in the next section.

1.2 Plurisubharmonic Functions

The starting point of this section lies in the observation that holomorpic functions are closely connected to harmonic and subharmonic functions. More precisely, the real part $u = \mathrm{Re}(f)$ of a holomorphic function $f : G \to \mathbb{C}$ defined on an open set G in the complex plane $\mathbb{C}$ is *harmonic*, i.e., its Laplacian vanishes,

$$\Delta u = 4 \frac{\partial^2 u}{\partial z \partial \bar{z}} = 0.$$

Moreover, $\log |f|$ is also harmonic on $G \setminus \{f = 0\}$, since $\Delta \log |f| = 0$, but the function $v = |f|$ is only *subharmonic* on G, i.e.,

$$\Delta v \geq 0 \tag{1.1}$$

Every harmonic function is C^∞-smooth (and real analytic), but for subharmonicity less regularity can be assumed. Either the Laplacian Δ is regarded in the sense of distributions or currents (cf. [104]) or it is defined via the following local maximum property.

Definition 1.2 Let $u : G \to [-\infty, \infty)$ be an upper semi-continuous function on a domain $G \subset \mathbb{C}$. Then u is called *subharmonic* if for every disc $\Delta \Subset G$ and every harmonic function h defined on some neighborhood of $\overline{\Delta}$ such that $u \leq h$ on $\partial \Delta$ we have that $u \leq h$ on the whole of $\overline{\Delta}$.

In higher dimensions, for a given holomorphic function $f : D \to \mathbb{C}$ defined on a domain $D \subset \mathbb{C}^n$, an easy computation yields the fact that $|f|$ and $\log |f|$ are subharmonic on $L \cap D$ for each complex line $L \simeq \mathbb{C}$ in $\mathbb{C}^n$. Functions fulfilling this condition determine a proper subset of the family of subharmonic functions. They were introduced in 1942 by Lelong [113, 114] ("fonctions plurisousharmoniques") and Oka [135] ("fonctions pseudoconvexes").

Definition 1.3 Let U be an open set in $\mathbb{C}^n$ and let $\psi : U \to [-\infty, \infty)$ be upper semi-continuous. Then ψ is called *plurisubharmonic* on U if and only if for every $p \in U$ and every direction $X \in \mathbb{C}^n$ with unit norm $|X| = 1$ the function $\zeta \mapsto \psi(p + \zeta X)$ is subharmonic. It is called *strictly plurisubharmonic* if for each $p \in U$ there exists an $\varepsilon > 0$ such that $\psi - \varepsilon |z|^2$ is still plurisubharmonic locally near p. If ψ and $-\psi$ both are plurisubharmonic, then ψ is called *pluriharmonic*.

In analogy to (1.1), by studying the local Taylor expansion, we obtain the characterization of plurisubharmonic functions in terms of their second derivatives.

Theorem 1.6 *Let U be an open set in $\mathbb{C}^n$ and let $\psi : U \to \mathbb{R}$ be a C^2-smooth function. Then ψ is plurisubharmonic (strictly plurisubharmonic) if and only if for every $p \in U$ the* complex Hessian *or* Levi matrix

$$\mathcal{L}_\psi(p) = \left(\frac{\partial^2 \psi}{\partial z_i \partial \overline{z}_j}(p) \right)_{i,j=1,\dots,n}$$

has only non-negative (only positive, resp.) eigenvalues. As a consequence, ψ is pluriharmonic if and only if its Levi matrix vanishes on U.

Every pluriharmonic function is locally the real part of a holomorphic function and, thus, it is real analytic and smooth (cf. [153]). Each plurisubharmonic function is subharmonic and, therefore, locally integrable. This implies that the $(-\infty)$-*locus* $\{\psi = -\infty\}$ of a plurisubharmonic function ψ has no interior points, if ψ is not identically $-\infty$.

Definition 1.4 A set S in $\mathbb{C}^n$ is called *pluripolar* if there exists a plurisubharmonic function ψ such that ψ is not identically $-\infty$ and $S \subset \{\psi = -\infty\}$.

Sets of such form are of special interest and were investigated, e.g., by Josefson [101] who showed that locally pluripolar sets are globally pluripolar. Moreover, these sets reappear in the context of extensions of positive closed currents beyond pluripolar sets (Skoda–El Mir theorem, cf. [56, 156, 157]). In the spirit of Hartogs' theorem (Theorem 1.3), another application of pluripolar sets is due to Shcherbina [155].

Theorem 1.7 *If the graph G_f of a continuous function $f : D \to \mathbb{C}$ defined on a domain $D \subset \mathbb{C}^n$ is pluripolar, then f is holomorphic.*

The converse holds true, since $G_f = \{(z, \zeta) : z \in D, \ \log |f(z) - \zeta| = -\infty\}$.

1.3   Pseudoconvex Domains

In this section, we study the relation of domains of holomorphy and plurisubharmonic functions. We will use the Hermitian form induced by the Levi matrix.

Definition 1.5 Let ψ be twice differentiable at a point p. For $X, Y \in \mathbb{C}^n$ the *Levi form* of ψ at p is the Hermitian form induced by the Levi matrix of ψ, i.e.,

$$\mathcal{L}_\psi(p)(X, Y) := \sum_{k,l=1}^{n} \frac{\partial^2 \psi}{\partial z_k \partial \overline{z}_l}(p) X_k \overline{Y}_l.$$

We continue with the following definition which determines generalized convexity in terms of a boundary condition.

Definition 1.6 Let Ω be an open set in $\mathbb{C}^n$. Then Ω is called *Levi pseudoconvex* (*strictly pseudoconvex*) if for every point p in $\partial\Omega$ there exist a neighborhood U of p and a C^2-smooth *local defining function* ϱ on U, i.e., $\Omega \cap U = \{z \in U : \varrho(z) < 0\}$ and $\nabla\varrho(p) \neq 0$, such that at each $p \in \partial\Omega$ the Levi form of ϱ at p is positive semi-definite (positive definite, resp.) on the *holomorphic tangent space*

$$H_p \partial\Omega = \{X \in \mathbb{C}^n : \partial\varrho(p)(X) = 0\}.$$

Levi [115] in 1910 observed that every domain of holomorphy Ω with C^2-smooth boundary is Levi pseudoconvex. For some period, the central question in several complex variables was whether or not the converse of the assertion of Levi holds true. It is called the *Levi problem* after Levi's paper [116] and was solved in 1942 by Oka [135] in $\mathbb{C}^2$. Later, independently in the early 50s, it was solved by Bremermann [24], Norguet [128] and Oka [137] in the general case $\mathbb{C}^n$ for $n \geq 2$. Another proof using PDE methods and the solution of the $\bar{\partial}$-equation is due to Hörmander [94, 95].

Now denote by δ_L the *Hartogs radius*, i.e., the boundary distance function on Ω along a fixed complex affine line $L \simeq \mathbb{C}$. Hartogs [84] showed that the function $-\log\delta_L(z)$ is subharmonic on $\Omega \cap L$. Since $\inf_L \delta_L$ equals the ordinary boundary distance function

$$\Omega \ni z \mapsto \operatorname{dist}(z, \partial\Omega) = \inf\{|z - w| : w \in \partial\Omega\},$$

Oka [135, 137] concluded that $z \mapsto -\log\operatorname{dist}(z, \partial\Omega)$ is plurisubharmonic on Ω, even if Ω is assumed to be only a domain of holomorphy. It is known as *Oka's Lemma*. A similar result was obtained at the same time by Lelong [114].

Altogether, the solution of the Levi problem may be stated as follows.

Theorem 1.8 *Let D be a domain in $\mathbb{C}^n$. Then D is a domain of holomorphy if and only if D is* pseudoconvex *in the sense that $-\log\operatorname{dist}(z, \partial D)$ is plurisubharmonic on D.*

Oka's work implies in particular that a domain over $\mathbb{C}^n$ is holomorphically convex if and only if it is equivalent to a connected component of the structure sheaf of $\mathbb{C}^n$ (a Riemann domain of holomorphy over $\mathbb{C}^n$). It turned out, from the work of Docquier and Grauert [54], that many variants of the notion of pseudoconvexity are equivalent to each other, as we have already indicated. Now the continuity principle in Theorem 1.1 implies that the sublevel sets $\{\psi < c\}$ of plurisubharmonic functions serve as easy instances of pseudoconvex sets or, equivalently, domains of holomorphy. One is given by the *Hartogs triangle* $\{(z, w) : |z| < |w| < 1\}$ which is a standard example of a pseudoconvex set in $\mathbb{C}^2$ with non-smooth boundary at the origin. As simple as it seems, it is used in many more examples to study phenomena related to plurisubharmonicity and pseudoconvexity (cf. [61]).

Let us continue with an application of the preceding results. First, observe that $\log|f|$ is plurisubharmonic whenever f is holomorphic, but not every plurisubharmonic function is of the form $\log|f|$ (cf. [25]). However, Bremermann

[26] developed a technique which permits us to retrieve holomorphicity out of plurisubharmonicity.

Theorem 1.9 *Let D be a domain of holomorphy and let ψ be a continuous plurisubharmonic function on D. Then for every compact set $K \subset D$ and $\varepsilon > 0$ there exist finitely many functions $f_j \in O(D)$ and positive integers n_j, $j = 1, \ldots, k$, such that*

$$\psi \le \max\{\frac{1}{n_j} \log |f_j| : j = 1, \ldots, k\} < \psi + \varepsilon \text{ on } K.$$

The *main idea of the proof* is to use holomorphicity of the *Hartogs domain*

$$\Omega = \{(z, w) \in D \times \mathbb{C} : |w| < e^{-\psi(z)}\}.$$

Since ψ is plurisubharmonic, the set Ω is pseudoconvex. By virtue of the solution of the Levi problem and the remark before Cartan and Thullen's result (Theorem 1.4), it is a domain of existence of a holomorphic function $h : \Omega \to \mathbb{C}$ which cannot exceed analytically the domain Ω near any of its boundary points. The approximating sequence is then derived by the Cauchy–Hadamard formula on the radius of convergence R_z of series of the form[2] $h(z, w) = \sum_\alpha c_\alpha(z) w^\alpha$ for fixed z and the relation $R_z = e^{-\psi(z)}$. $\qquad\qquad\qquad\qquad\qquad\qquad\qquad\qquad\qquad\qquad\qquad\square$

As a consequence, there is another way to express holomorphically convex hulls from Theorem 1.4 in terms of plurisubharmonic functions.

Corollary 1.1 *Let D be a domain of holomorphy. Then for every compact set K in D we have the identity*

$$\widehat{K}_D = \{z \in D : \psi(z) \le \max_K \psi \ \forall \ \psi \text{ plurisubharmonic on } D\}.$$

Since holomorphic and plurisubharmonic functions are locally defined, they easily transfer to complex manifolds (and Riemann domains). Those complex manifolds which are holomorphically convex and separate points by holomorphic functions are known as *Stein manifolds* introduced by Karl Stein [161] in 1951. Here, a *point separating manifold M* is defined by the property that for each two points $p, q \in M$ with $p \ne q$ there exists a function $f \in O(M)$ such that $f(p) \ne f(q)$. Stein manifolds can be described in terms of plurisubharmonic functions as follows [54].

Theorem 1.10 (Docquier–Grauert) *A complex manifold M is Stein if and only if it admits a continuous strictly plurisubharmonic exhaustion function ψ on M, i.e., $\{\psi < c\}$ is relatively compact in M for each $c \in \mathbb{R}$.*

---

[2] Notice that the convergence of this type of series were already studied earlier by Hartogs [82].

Now if ψ is a smooth strictly plurisubharmonic exhaustion function, the positive closed $(1, 1)$-form $\omega = i\partial\bar{\partial}\psi$ induces a complete Kähler metric on M. Therefore, any Stein manifold M is complete Kähler. But the converse is not true in general. For any closed analytic subset of a Stein manifold, the set $M \setminus A$ is complete Kähler, but if the codimension of A exceeds 2, then $M \setminus A$ cannot be Stein by Remmert's theorem on removable singularities. These statements were already verified in Grauert's thesis from 1956.

In view of Hartogs' theorem, the following result was shown in [130] for the C^1-smooth and, recently, in [133] for the continuous case.

Theorem 1.11 *A continuous function $f : \Delta^n \to \mathbb{C}$ is holomorphic if and only if the complement of its graph G_f in $\Delta^n \times \mathbb{C}$ carries a complete Kähler metric.*

The subsequent characterization of Stein manifolds was already established basically by Oka in [136], but is often attributed to Cartan [32] (Theorem B).

Theorem 1.12 (Oka–Cartan Fundamental Theorem on Stein Manifolds) *If M is a Stein manifold, then for every coherent analytic sheaf $\mathcal{F}$ and $k \geq 1$ the k-th cohomology group $H^k(M, \mathcal{F})$ vanishes.*

R. Narasimhan [123] and Grauert [75] solved the Levi problem on complex spaces resulting in the notion of *Stein spaces*, and Cartan's Theorem B transfers to Stein spaces (cf. [78]). Complex spaces are, roughly speaking, locally biholomorphic to the zero level set of finitely many analytic functions, or can be interpreted as complex manifolds with singularities. Those are *Stein* if they admit a smooth strictly plurisubharmonic exhaustion function. Thanks to Richberg's results on the approximation of plurisubharmonic functions [149], it suffices to assume only continuity of the exhaustion function.

Fornæss and Narasimhan [60] proved the equivalence of plurisubharmonicity and weak plurisubharmonicity, which is defined as follows.

Definition 1.7 An upper semi-continuous function $\psi : X \to [-\infty, \infty)$ defined on a (reduced) complex space X is called *weakly plurisubharmonic* if for each holomorphic disc $f : \Delta \to X$ the composition $\psi \circ f$ is subharmonic on Δ.

This characterization is useful for constructing a plurisubharmonic *upper envelope function* in order to solve the Dirichlet problem (e.g. [146]).[3] Furthermore, it allows us to introduce plurisubharmonicity in almost complex manifolds due to the existence of local discs [52].

We finish this chapter with a brief remark on the compact case. It is well known that the complex manifolds struggle with holomorphic functions in the sense that there are only constant ones by virtue of Liouville's theorem. The same holds true for plurisubharmonic functions. Therefore, compactness seems to be contrary to

---

[3] Since the solution of the Dirichlet problem leads to another Pandora's box known as *pluripotential theory*, we will not dig deeper into this interesting topic. The interested reader may enjoy, for instance, Klimek's book [104].

Steinness in a certain sense. Nevertheless, Cartan and Serre [34] proved a statement in the spirit of Oka and Cartan in the compact case.

Theorem 1.13 (Cartan–Serre Finiteness Theorem) *If X is a compact complex space, then for every coherent analytic sheaf $\mathcal{F}$ and $k \geq 1$ the cohomology group $H^k(X, \mathcal{F})$ is finite.*

We will encounter another finiteness theorem for q-convex domains later in Chap. 4, but before then, we continue with the discussion on generalized plurisubharmonicity and pseudoconvexity, namely, *q-plurisubharmonicity* and *q-pseudoconvexity*.

Chapter 2
Basics of q-Plurisubharmonic Functions

This chapter is partially extracted from the doctoral thesis[1] [139] and the collaborating papers [141–143] of the second-named author together with E. S. Zeron. For this reason, it contains both classical and the author's recent results on the topic of q-plurisubharmonic functions. It serves as a preparation for Chap. 3 in which we study domains created by *q-plurisubharmonic* functions. These were introduced by Hunt and Murray [97] in 1978 who defined them in terms of a local maximum property, similar to subharmonicity, but replacing harmonic functions by functions pluriharmonic on $(q + 1)$-dimensional subspaces. This means that 0-plurisubharmonic functions form the classical plurisubharmonic functions. In the smooth case, q-plurisubharmonic functions are the same as (weakly) $(q + 1)$-*convex* functions in the sense of Grauert [73] about which we learn more in Chap. 4.

We collect the main properties of q-plurisubharmonic functions in a more or less chronological order by their appearance during previous decades starting from the above-mentioned introduction by Hunt and Murray, with important developments by Słodkowski [159, 160] and further results by O. Fujita [69], Bungart [29] and Nguyen Quang Dieu [53]. Especially, we recall approximation techniques for q-plurisubharmonic functions established by Słodkowski [159] and Bungart [29]. At the end, we devote a section to q-*holomorphic* functions in the sense of Basener [18] and their relation to q-plurisubharmonicity.

---

[1] The initial idea of the study of q-plurisubharmonicity was to generalize classical results in complex analysis to the setting of generalized convexity, e.g., to solve the Dirichlet problem for q-plurisubharmonic functions on bounded (or unbounded) domains. Eventually, the thesis evolved in a compendium on q-plurisubharmonicity together with various results in this field, such as the study of the Shilov boundary for q-plurisubharmonic functions, convex hulls created by q-plurisubharmonic functions, and complex foliations of complements of q-pseudoconvex sets.

© The Author(s), under exclusive license to Springer Nature Singapore Pte Ltd. 2022
T. Ohsawa, T. Pawlaschyk, *Analytic Continuation and q-Convexity*, SpringerBriefs
in Mathematics, https://doi.org/10.1007/978-981-19-1239-9_2

2.1   Basic Properties of q-Plurisubharmonic Functions

The q-plurisubharmonic functions originate in the article [97] by Hunt and Murray from 1978. They defined a C^2-smooth function to be *q-plurisubharmonic* on a domain U in $\mathbb{C}^n$ if its Levi matrix has at least $(n - q)$ non-negative eigenvalues at each point p in U. Those who are familiar with this topic immediately recognize this condition as determining (weakly) $(q + 1)$-convexity in the sense of Grauert [73].[2] We will devote the whole of Chap. 4 to q-convexity later. Hunt and Murray pose the question whether there is a non-smooth analogous characterization of q-plurisubharmonicity. They suggested the subsequent definition.

Definition 2.1 Let $q \in \{0, \dots, n - 1\}$ and let $\psi : \Omega \to [-\infty, \infty)$ be *an upper semi-continuous* function on an open set Ω in $\mathbb{C}^n$, i.e., for every $c \in \mathbb{R}$ the sublevel set $\{\psi < c\}$ is open.

1. The function ψ is called *$(n - 1)$-plurisubharmonic on Ω* if for every ball $B \Subset \Omega$ and every function h which is pluriharmonic on a neighborhood of $\overline{B}$ with $\psi \leq h$ on ∂B we already have that $\psi \leq h$ on $\overline{B}$.
2. The function ψ is *q-plurisubharmonic* on Ω if ψ is q-plurisubharmonic on $\pi \cap \Omega$ for every complex affine subspace π of dimension $q + 1$.
3. If $q \geq n$, every upper semi-continuous function on Ω is by convention q-plurisubharmonic.
4. An upper semi-continuous function ψ on Ω is called *strictly q-plurisubharmonic on Ω* if for every point p in Ω there is an $\varepsilon > 0$ such that $\psi - \varepsilon|z|^2$ remains q-plurisubharmonic near p.

Instead of using the notion of $(n - 1)$-*plurisubharmonic*, we rather prefer to call this type of functions *subpluriharmonic*, since we feel it fits better to its characterizing property, as in the case of subharmonicity (cf. Definition 1.2).[3]

According to the above definition, the 0-plurisubharmonic functions are exactly the plurisubharmonic functions. Moreover, it does not matter whether subpluriharmonic functions are defined via pluriharmonic, smooth plurisuperharmonic, upper semi-continuous plurisuperharmonic or even real parts of holomorphic polynomials due to appropriate approximation techniques for plurisubharmonic functions (cf. Richberg's approximation [149] and Theorem 1.9), the relation of pluriharmonic and holomorphic functions and the fact that $(n - 1)$-plurisubharmonicity is a local property (cf. [160]).

---

[2] Oddly, Grauert remained unmentioned in Hunt and Murray's article even though the theory of q-convexity was already widely explored at that time.

[3] When searching for q-plurisubharmonicity, one rapidly encounters *q-(pluri)subharmonic functions* in the sense of Ho Lop–Hing [92]. It differs from the definition of Hunt and Murray by using harmonic instead of pluriharmonic functions on q-dimensional complex subspaces, as the name indicates. For this reason, it seems that *q-subpluriharmonic* constitutes a more appropriate name for q-plurisubharmonic functions, but in order to remain consistent with the existing literature (and to show respect to the originators), we keep using the notion of q-plurisubharmonicity.

We give a list of properties of q-plurisubharmonic functions which easily follow from the definition.

Proposition 2.1 *Every below-mentioned function is defined on an open set Ω in $\mathbb{C}^n$ unless otherwise stated.*

1. *Every q-plurisubharmonic function is $(q + 1)$-plurisubharmonic.*
2. *The limit of a decreasing sequence $(\psi_k)_{k \in \mathbb{N}}$ of q-plurisubharmonic functions is again q-plurisubharmonic.*
3. *If $\{\psi_i : i \in I\}$ is a family of locally bounded above q-plurisubharmonic functions, then the upper semi-continuous regularization*

$$\psi^*(z) := \limsup_{\substack{w \to z \\ w \neq z}} \psi(w)$$

   of $\psi := \sup_{i \in I} \psi_i$ is q-plurisubharmonic.
4. *An upper semi-continuous function ψ is q-plurisubharmonic on Ω if and only if it is locally q-plurisubharmonic on Ω, i.e., for each point p in Ω there is a neighborhood U of p in Ω such that ψ is q-plurisubharmonic on U.*
5. *Let ψ be a q-plurisubharmonic function on Ω and $F : \mathbb{C}^n \to \mathbb{C}^n$ a $\mathbb{C}$-linear isomorphism. Then the composition $\psi \circ F$ is q-plurisubharmonic on $\Omega' := F^{-1}(\Omega)$. Moreover, if ψ is subpluriharmonic on Ω and F is biholomorphic from an open set G in $\mathbb{C}^n$ onto Ω, then $\psi \circ F$ is subpluriharmonic on G.*
6. *A function ψ is q-plurisubharmonic on Ω if and only if for every ball $B \Subset \Omega$ and for every plurisubharmonic function s on B in Ω the sum $\psi + s$ is q-plurisubharmonic on B.*

An important tool is to glue q-plurisubharmonic functions in order to get a new one on a larger domain (cf. [97]).

Lemma 2.1 *Let Ω_1 be an open set in Ω. Let ψ be a q-plurisubharmonic function on Ω and ψ_1 be a q-plurisubharmonic function on Ω_1 such that*

$$\limsup_{\substack{w \to z \\ w \in \Omega_1}} \psi_1(w) \le \psi(z) \ \text{for every } z \in \partial\Omega_1 \cap \Omega.$$

Then the subsequent function is q-plurisubharmonic on Ω,

$$\varphi(z) := \left\{ \begin{array}{ll} \max\{\psi(z), \psi_1(z)\}, & \text{if } z \in \Omega_1 \\ \psi(z), & \text{if } z \in \Omega \setminus \Omega_1 \end{array} \right\}.$$

It follows directly from the definition that, for $q < n$, a q-plurisubharmonic function defined on a neighborhood of the closure of a bounded domain Ω in $\mathbb{C}^n$ attains its maximum on the boundary of Ω. Hunt and Murray [97] generalized this observation as follows.

Theorem 2.1 (Local Maximum Property) *Let $q \in \{0, \ldots, n-1\}$ and Ω be a relatively compact open set in $\mathbb{C}^n$. Then any function u which is upper semi-continuous on $\overline{\Omega}$ and q-plurisubharmonic on Ω fulfills*

$$\max\{\psi(z) : z \in \overline{\Omega}\} = \max\{\psi(z) : z \in \partial\Omega\}.$$

By carefully investigating the Taylor expansion, Hunt and Murray [97, Lemma 2.6] gave a characterization of smooth q-plurisubharmonic functions in terms of the number of negative eigenvalues of the Levi matrix.

Then q-plurisubharmonicity can be determined by its Levi matrix similarly to the plurisubharmonic case (cf. Theorem 1.6).

Theorem 2.2 *Let $q \in \{0, \ldots, n-1\}$ and let ψ be a C^2-smooth function on an open subset Ω in $\mathbb{C}^n$. Then ψ is q-plurisubharmonic (strictly q-plurisubharmonic) if and only if the Levi matrix has at most q negative (q non-positive, resp.) eigenvalues at every point p in Ω.*

By virtue of this characterization, smooth q-plurisubharmonic and strictly q-plurisubharmonic functions are exactly the weakly $(q + 1)$-convex and $(q + 1)$-convex functions, respectively, in the sense of Grauert [73]. Moreover, one immediately derives the subsequent important examples which reveal the main difference between plurisubharmonicity and q-plurisubharmonicity for $q \geq 1$.

Example 2.1 The family of q-plurisubharmonic functions is not stable under summation. A simple example in $\mathbb{C}^2$ is given by the 1-plurisubharmonic functions $\psi_1(z, w) = -|z|^2$ and $\psi_2(z, w) = -|w|^2$, and their sum $\psi(z, w) = -|z|^2 - |w|^2$, which is only 2-plurisubharmonic.

Example 2.2 The q-plurisubharmonic functions are not locally integrable in general. Indeed, for $k \in \mathbb{N}$ consider the 1-plurisubharmonic function $f_k(z, w) := -k|z|^2$ in $\mathbb{C}^2$. Then the sequence $(f_k)_{k \in \mathbb{N}}$ decreases to the 1-plurisubharmonic *characteristic function* χ_A of the line $A = \{z = 0\}$,

$$\chi_A := \begin{cases} 0, & z \in A \\ -\infty, & z \notin A \end{cases}.$$

Since the latter function is equal to $-\infty$ at inner points outside A in $\mathbb{C}^2$, it fails to be locally integrable, whereas any plurisubharmonic function lies in L^1_{loc}, except for the constant function $\varphi = -\infty$.

Example 2.3 Every entire plurisubharmonic function on $\mathbb{C}^n$ admits the *Liouville property*, i.e., it is constant whenever it is bounded from above on $\mathbb{C}^n$. But the previous example demonstrates that it is no longer true for q-plurisubharmonic functions if $q \geq 1$. Another example is given by the function $\psi(z, w) = -1/(1+|z|^2+|w|^2)$. It is strictly 1-plurisubharmonic on $\mathbb{C}^2$ and upper bounded, but not constant. By a similar reasoning, there exist non-constant 1-plurisubharmonic functions on the compact complex manifold $\mathbb{CP}^2$.

Example 2.4 The function $(z, w) \mapsto |zw|^2$ is strictly 1-plurisubharmonic but fails to be strictly plurisubharmonic on $\mathbb{C}^2$. Indeed, the eigenvalues of its Levi matrix at a given point (z, w) are zero and $|z|^2 + |w|^2$. Nonetheless, it is plurisubharmonic.

Example 2.5 If $f : \Omega \to \mathbb{C}^m$ is a holomorphic mapping on an open set Ω in $\mathbb{C}^n$, then $-\log \|f\|$ and $1/\|f\|$ are subpluriharmonic outside $\{f = 0\}$ for *any* complex norm $\| \cdot \|$ in $\mathbb{C}^m$ (cf. [142]). As a consequence, if

$$d_{\|\cdot\|}(z, \partial\Omega) := \inf\{\|z - w\| : w \in \partial\Omega\}$$

is the boundary distance function induced by $\| \cdot \|$ between z in Ω and $\partial\Omega$, then

$$z \to -\log d_{\|\cdot\|}(z, \partial\Omega)$$

is subpluriharmonic on Ω. As revealed by Słodkowski [160], this means that every domain Ω in $\mathbb{C}^n$ admits an $(n - 1)$-plurisubharmonic exhaustion function. In this sense, it is $(n-1)$-*pseudoconvex*. This fact follows already from Green–Wu's result in [80]. We will investigate q-pseudoconvex domains later in Chap. 3.

2.2 Approximation of q-Plurisubharmonic Functions

Hunt and Murray [97] intended to transfer results by Bremermann [27] and Walsh [167] on the Dirichlet problem for plurisubharmonic functions defined on strictly pseudoconvex sets to q-plurisubharmonicity. Therefore, they introduced generalized (Levi) pseudoconvex sets (cf. Definition 1.6).

Definition 2.2 An open set Ω in $\mathbb{C}^n$ is called *Levi q-pseudoconvex* or *strictly (Levi) q-pseudoconvex* if for every point p in $\partial\Omega$ there exist a neighborhood U of p and a C^2-smooth local defining function ϱ such that its Levi form $\mathcal{L}_\varrho$ at p has at most q negative or, respectively, at most q non-positive eigenvalues on the holomorphic tangent space $H_p\partial\Omega$.

They were able to show that, given a continuous function b on the boundary of a bounded strictly q-pseudoconvex domain Ω, the *upper envelope function*

$$\psi : z \mapsto \sup\{u(z) : u \ q\text{-plurisubharmonic on } \overline{\Omega}, \ u \le b \text{ on } \partial\Omega\}$$

is continuous, q-plurisubharmonic and, additionally, that $-\psi$ is $(n - q - 1)$-plurisubharmonic. Here, by q-plurisubharmonicity of u on $\overline{\Omega}$ we mean that it extends to an upper semi-continuous function up to the boundary of Ω. Now if we assume that ψ is C^2-smooth, then its Levi matrix would have at least one zero eigenvalue. But this means that ψ solves the homogeneous complex Monge–Ampère equation $(\partial\overline{\partial}\psi)^n = 0$. Under the extra condition $2q < n$ they verified that the upper envelope function attains the boundary value b at $\partial\Omega$. Thus, it solves the Dirichlet problem for q-plurisubharmonic functions. They conjectured that the solution is unique, but were able to verify their assertion only in the pseudoconvex case $q = 0$. Kalka [102] proved that their conjecture holds true, but restricted the definition of the upper envelope to a smaller subfamily of q-plurisubharmonic functions. It was Słodkowski [159] in 1984 who gave a full positive answer to their conjecture. First, he proved that the sum of a q-plurisubharmonic and an r-plurisubharmonic function is indeed $(q + r)$-plurisubharmonic. Now if we assume that there exists another solution of the Dirichlet problem fulfilling the properties of the upper envelope ψ, say φ, then we have that φ is q-plurisubharmonic and $-\varphi$ is $(n - q - 1)$-plurisubharmonic on $\overline{\Omega}$. By Słodkowski, both the functions $\psi - \varphi$ and $\varphi - \psi$ are $(n - 1)$-plurisubharmonic on Ω and vanish on $\partial\Omega$. By applying the local maximum principle, it follows that $\psi - \varphi$ has to vanish everywhere on Ω. Hence, φ agrees with the upper envelope ψ.

While the proof of Słodkowski's statement on the sum of q- and r-plurisubharmonic functions is an easy exercise in the smooth case, the general upper semi-continuous case demands advanced techniques from the theory of convex functions, in particular, the classical result by Alexandroff and Busemann-Feller [1, 31]. It states that every real-valued convex function is in fact twice differentiable almost everywhere. As a consequence, each locally *semi-convex* function is twice differentiable almost everywhere. These are functions u such that $u + L|z|^2$ is locally convex for some real number L. Now let ψ be a q-plurisubharmonic function on an open set Ω in $\mathbb{C}^n$ and denote by $\widehat{\psi}$ the upper semi-continuous extension of ψ by $-\infty$ into the whole of $\mathbb{C}^n$. Consider the *supremum convolution* $\psi *_s \theta$ of ψ with a smooth function θ with compact support in Ω defined by

$$(\psi *_s \theta)(z) := \sup\{\widehat{\psi}(w)\theta(z - w) : w \in \mathbb{C}^n\}.$$

Then $\psi *_s \theta$ is locally semi-convex, continuous, almost everywhere twice differentiable and q-plurisubharmonic on a neighborhood of some compact set inside Ω. Now if the volume of the support of θ tends to zero, $\psi *_s \theta$ converges pointwise from above towards ψ on a prescribed compact set inside Ω. With some effort, Słodkowski managed to establish a characterization similar to Theorem 2.2 in the case of functions which are only twice differentiable almost everywhere. Together with the above-described approximation technique, he showed the following statement [159, Theorem 4.1].

Theorem 2.3 *Let ψ be a q-plurisubharmonic function and φ an r-plurisubharmonic function both defined on Ω. Then $\psi + \varphi$ is $(q + r)$-plurisubharmonic.*

Another application of methods from convexity, the so-called *viscosity subsolutions*, is involved in order to solve generalized boundary value problems with the Perron method [46]. In regards to q-plurisubharmonicity, these were studied by Harvey and Lawson [88] in the case $q = 0$ and in [143] in the case $q \geq 1$.

Theorem 2.4 *Let $q \geq 0$ be any integer, and $\Omega \subset \mathbb{C}^n$ be an open set. An upper semi-continuous function $u : \Omega \to [-\infty, \infty)$ is q-plurisubharmonic on Ω if and only if it is q-plurisubharmonic in the viscosity sense on Ω, i.e., for every $\varphi \in C^2(\Omega)$*

the Levi matrix of φ has at most q negative eigenvalues at those points where $u - \varphi$ attains its maximum.

Due to an example of Diederich and Fornæss (cf. [50, Theorem 2]), it is not possible to approximate q-plurisubharmonic functions from above by a sequence of C^2-smooth ones in general. Anyway, Bungart [29] proved that there is an approximation of continuous q-plurisubharmonic functions by piecewise smooth ones.

Definition 2.3 Let ψ be a continuous function on an open set Ω in $\mathbb{C}^n$. The function ψ is called *q-plurisubharmonic with corners* if for every point p in Ω there is an open neighborhood U of p in Ω and finitely many C^2-smooth q-plurisubharmonic functions $\psi_1, \ldots, \psi_k$ on U such that $\psi = \max\{\psi_j : j = 1, \ldots, k\}$.

Bungart's approximation method is based on the solution of the Dirichlet problem and the Perron method. The main idea is to approximate a given continuous q-plurisubharmonic ψ from above by arbitrary continuous ones, say $\{f_\nu\}_\nu$. Fix some f_ν close enough to ψ. Then we cover the given domain Ω from inside by open strictly pseudoconvex sets, such as balls $\{B_j\}_j$, and take the upper envelope function $P_{\nu,j}$ on each of these balls induced by the family of q-plurisubharmonic functions with corners and with prescribed values f_ν on ∂B_j. Finally, if we carefully glue together the $P_{\nu,j}$'s, we end up with a piecewise smooth q-plurisubharmonic function close enough to ψ.

Theorem 2.5 *Let Ω be an open subset in $\mathbb{C}^n$ and let ψ be continuous and q-plurisubharmonic on Ω. Then for every continuous function f on Ω with $\psi < f$ there exists a q-plurisubharmonic function φ with corners on Ω such that $\psi \leq \varphi < f$.*

Combining Słodkowski's and Bungart's approximation methods, we obtain that, for a given q-plurisubharmonic function on Ω and a compact set K in Ω, there exists a sequence of q-plurisubharmonic functions ψ_j with corners decreasing to ψ on K (cf. [53]). These approximation techniques will be applied in the next section.

2.3 Weakly q-Plurisubharmonic Functions and Compositions

We present results on compositions of mappings in which q-plurisubharmonic functions are involved. In the smooth case, a simple computation of the Levi matrix immediately yields the subsequent statement (cf. [53]). In the non-smooth case, it is due to the approximation techniques by Słodkowski and Bungart described in the previous section.

Theorem 2.6 *Let ω be an open set in $\mathbb{R}$ and let u be a real-valued, non-decreasing convex function on ω. Let ψ be q-plurisubharmonic on an open set Ω in $\mathbb{C}^n$ and assume that $\psi(\Omega)$ lies in ω. Then the composition $u \circ \psi$ is q-plurisubharmonic.*

Similarly, the composition $u \circ \psi$ of a q-plurisubharmonic function and an arbitrary real-valued function u always produces a $(q + 1)$-plurisubharmonic function.

Notice that it follows directly from the definition that $\psi \circ F$ stays q-plurisubharmonic whenever F is a linear isomorphism. In fact, a more general statement holds true and was proved by O. Fujita [69].

Theorem 2.7 *A function ψ is q-plurisubharmonic on an open set Ω in $\mathbb{C}^n$ if and only if it is* weakly *q-plurisubharmonic, i.e., $\psi \circ f$ is q-plurisubharmonic for every holomorphic mapping $f : D \to \Omega$, where D is a domain in $\mathbb{C}^{q+1}$.*

Fujita's idea was to show that the Hartogs domain

$$\Omega' = \{(z, w) \in \Omega \times \mathbb{C}, \ |w| < e^{-\psi(z)}\}$$

is *pseudoconvex of order $n - q$* in $\mathbb{C}^{n+1}$ if and only if

$$\psi(z) = -\log \operatorname{dist}\big((z, 0), \partial\Omega' \cap (\{z\} \times \mathbb{C}_w)\big)$$

is (weakly) q-plurisubharmonic on Ω. Notice that Fujita used the notions *pseudoconvexe d'ordre général* and *pseudoconvexe d'ordre $n - q$*. We will learn more about generalized pseudoconvexity in Chap. 3.

Later, Nguyen Quang Dieu [53] presented a proof which only involves q-plurisubharmonic functions and the approximation techniques by Słodkowski and Bungart. Using them, the problem reduces to counting the eigenvalues of the Levi matrix of the composition $\psi \circ f$ of a twice differentiable q-plurisubharmonic function ψ with a holomorphic mapping f.

As an application, we learn more on characteristic functions (cf. Example 2.2).

Example 2.6 Let Ω be an open set in $\mathbb{C}^n$ and let $h : \Omega \to \mathbb{C}^q$ be a holomorphic mapping. Let $| \cdot |$ be the complex Euclidean norm on $\mathbb{C}^q$ and consider the analytic variety $A = \{h = 0\}$. Then for each $k \in \mathbb{N}$ the function $\chi_k(z) := -k|h(z)|^2$ is q-plurisubharmonic on Ω and decreases to the characteristic function

$$\chi_A(z) = \begin{cases} 0, & \text{if } z \in A \\ -\infty, & \text{if } z \in \Omega \setminus A \end{cases}.$$

By Proposition 2.1, point 2, the function χ_A is q-plurisubharmonic on Ω. We will generalize this observation to analytic sets in Thoerem 2.10.

The previous discussion allows us to introduce q-plurisubharmonicity on complex manifolds, but there is not much to extract from the literature (except e.g. [117, 118]). In fact, it is the q-*convexity* which has been studied intensively in complex manifolds and complex spaces (see Chap. 4). Nevertheless, by courtesy of O. Fujita, there are (at least) two ways to introduce q-plurisubharmonicity on analytic sets.

Definition 2.4 Let A be an analytic subset of an open set Ω in $\mathbb{C}^n$. Let ψ be an upper semi-continuous function on A.

1. We say that ψ is q-*plurisubharmonic* on A if for every point p in A there exist an open neighborhood U of p in Ω and a q-plurisubharmonic function Ψ on U such that $\psi = \Psi$ on $A \cap U$.
2. The function ψ is *weakly q-plurisubharmonic* on A if for every holomorphic mapping $f : D \to A$ defined on a domain D in $\mathbb{C}^{q+1}$ the composition $\psi \circ f$ is q-plurisubharmonic on D.

Fornæss and Narasimhan showed that these two notions are equivalent in the plurisubharmonic case $q = 0$ (cf. [60]). Later, Popa-Fischer [147] generalized their result to continuous q-plurisubharmonic functions. In fact, the following statement holds true in reduced complex spaces, but the following statement suffices for our purposes.

Theorem 2.8 *A continuous function on an analytic subset A is weakly q-plurisubharmonic if and only if it is q-plurisubharmonic.*

Anyhow, it is still an open question whether an upper semi-continuous weakly q-plurisubharmonic function on an analytic set is q-plurisubharmonic.

It is well-known that, if Ω is pseudoconvex, a function which is holomorphic on an analytic subset A on Ω extends holomorphically to the whole of Ω (cf. [78]). From this, Colţoiu achieved, with some effort, a similar extension theorem for plurisubharmonic functions (cf. Proposition 2 in [39]).

Theorem 2.9 *Let A be an analytic subset of a Stein open set Ω in $\mathbb{C}^n$. Then every plurisubharmonic function ψ on A extends to a plurisubharmonic function Ψ into the whole of Ω. Moreover, if K is a compact holomorphically convex set in Ω and $\psi < 0$ on $A \cap K$, then $\Psi < 0$ on K.*

It is still an open question whether his result carries over to the case $q \geq 1$. Anyway, we present an approach to this issue (cf. [139]). Recall that, given a point p in A, the *dimension of A at p* is defined by

$$\dim_p A := \limsup_{A^* \ni z \to p} \dim_z A,$$

where A^* denotes the set of all regular points of A and $\dim_z A$ is the dimension of the complex manifold $A \cap U$ in some neighborhood U of the regular point $z \in A^*$. For more details on analytic sets we recommend Chirka's book [37].

Theorem 2.10 *Let $q \in \{0, \ldots, n-1\}$ and let $r \geq 0$ be an integer. Let Ω be an open set in $\mathbb{C}^n$ and let A be an analytic subset of Ω. Let*

$$q = q_A := n - \min\{\dim_z A : z \in A\}.$$

If the function ψ is r-plurisubharmonic on A, then ψ extends to a $(q + r)$-plurisub-harmonic function Ψ_A on the whole of Ω via the trivial extension

$$\Psi_A(z) := \left\{ \begin{array}{ll} \psi(z), & \text{if } z \in A \\ -\infty, & \text{if } z \in \Omega \setminus A \end{array} \right\}.$$

Now if Ω is a union of analytic subsets $(A_j)_{j \in J}$ with constant $q = q_{A_j}$ for each j, e.g., coming from a complex foliation of Ω, and Ψ is an upper semicontinuous function on Ω such that Ψ is r-plurisubharmonic on each A_j, then $\Psi = \sup\{\Psi_{A_j} : j \in J\}$ is $(q + r)$-plurisubharmonic on Ω. Notice that this observation is relevant in the plurisubharmonic case $r = 0$ and implies that $-\log \|\cdot\|$ is subpluriharmonic on $\mathbb{C}^n \setminus \{0\}$ for *any* complex norm $\|\cdot\|$ (cf. Example 2.5).

Słodkowski transferred the local maximum property to q-plurisubharmonic functions on analytic sets (cf. Proposition 5.2 and Corollary 5.3 in [160]). Here, it is an immediate consequence of Theorem 2.10.

Theorem 2.11 (Local Maximum Property on Analytic Sets) *Let $q \in \{0, \ldots, n - 1\}$ and let A be an analytic subset of an open set Ω in $\mathbb{C}^n$ such that $\dim_z A \geq q + 1$ for every $z \in A$, and let ψ be a q-plurisubharmonic function on A. Then for every compact set K in A we have that*

$$\max\{\psi(z) : z \in K\} = \max\{\psi(z) : z \in \partial_A K\}.$$

Here, by $\partial_A K$ we mean the relative boundary of K in A.

Indeed, according to the previous theorem, the trivial extension Ψ_A is $(n - 1)$-plurisubharmonic on Ω, so the local maximum principle on $\mathbb{C}^n$ applies (Proposition 2.1). This tool will turn out to be important when we deal with a version of the continuity principle for q-pseudoconvex sets.

2.4 q-Holomorphic Functions

It is easy to compute that $\log |f|$ and $|f|$ are both plurisubharmonic if f is a holomorphic function. Now it is natural to ask what is the appropriate condition for a smooth function $f : \Omega \to \mathbb{C}$ so that $\log |f|$ is q-plurisubharmonic. An answer was given by Hunt and Murray [97] by using smooth functions $f : \Omega \to \mathbb{C}$ fulfilling the non-linear Cauchy-Riemann equation

$$\overline{\partial} f \wedge (\partial \overline{\partial} f)^q = 0,$$

the so-called q-*holomorphic* functions. Initially in 1976, Basener [18] investigated q-holomorphic functions and their relation to q-pseudoconvex domains while, later, Hunt and Murray used Basener's results to create q-plurisubharmonic peak functions out of q-holomorphic ones at strictly q-pseudoconvex boundary points.

From Basener's article, we extract a collection of properties of q-holomorphic functions.

Proposition 2.2 *Let Ω be an open set in $\mathbb{C}^n$ and let $q \geq 0$ be an integer. All below-mentioned functions are assumed to be C^2-smooth and defined on Ω unless otherwise stated.*

1. *The 0-holomorphic functions are exactly the classical holomorphic ones. If $N \geq n$, by N-holomorphic functions we simply mean the C^2-smooth functions by convention.*
2. *Every q-holomorphic function is $(q + 1)$-holomorphic.*
3. *Let $r \geq 0$ be another integer. If f is q-holomorphic and g is r-holomorphic, their sum $f + g$ and the product fg are both $(q + r)$-holomorphic.*
4. *Let f be a q-holomorphic function on Ω. If $g : D \rightarrow \Omega$ is a holomorphic mapping defined on another open set D in $\mathbb{C}^k$, then the composition $f \circ g$ is q-holomorphic on D.*
5. *Given a q-holomorphic function f on Ω and a holomorphic function h defined on a neighborhood of the image of f in $\mathbb{C}$, then the composition $h \circ f$ is q-holomorphic on Ω. In particular, the power f^m is again q-holomorphic on Ω for every $m \in \mathbb{N}_0$.*

Basener presented another characterization in terms of the complex gradient and the Levi matrix of f.

Theorem 2.12 *The function f is q-holomorphic on Ω if and only if at each point in Ω the rank of the matrix*

$$\begin{pmatrix} f_{\bar{z}_1} & \cdots & f_{\bar{z}_n} \\ f_{z_1\bar{z}_1} & \cdots & f_{z_1\bar{z}_n} \\ \vdots & \ddots & \vdots \\ f_{z_n\bar{z}_1} & \cdots & f_{z_n\bar{z}_n} \end{pmatrix}$$

is less than or equal to q. Here, $f_z = \frac{\partial f}{\partial z}$, $f_{\bar{z}} = \frac{\partial f}{\partial \bar{z}}$ and $f_{z\bar{w}} = \frac{\partial^2 f}{\partial z \partial \bar{w}}$.

From this, Hunt and Murray derived the relation of q-holomorphic functions and q-plurisubharmonic ones (cf. Theorem 5.4 and Corollary 5.4 in [97]).

Theorem 2.13 *If f is q-holomorphic on Ω, then the real part $\operatorname{Re}(f)$, the imaginary part $\operatorname{Im}(f)$, the absolute value $|f|$ and $\log|f|$ are all q-plurisubharmonic.*

Therefore, the q-holomorphic functions inherit the local maximum modulus property from q-plurisubharmonicity.

Theorem 2.14 (Local Maximum Modulus Property) *Let $q \in \{0, \ldots, n - 1\}$ and let Ω be an open bounded set in $\mathbb{C}^n$. If f is q-holomorphic on Ω and continuous up* *to its boundary $\partial\Omega$, we have that*

$$\max\{|f(z)| : z \in \Omega\} = \max\{|f(z)| : z \in \partial\Omega\}.$$

It is natural to seek for a holomorphic structure inside a q-holomorphic function if $q \geq 1$. The following examples in [18] demonstrate the difficulty of this problem and underline the main difference between the holomorphic case $q = 0$ and the general case $q \geq 1$.

Example 2.7 If f is holomorphic, then the complex conjugate $\overline{f}$ and the absolute value $|f|^2$ are both 1-holomorphic. Since the real part of f is pluriharmonic, it is 1-holomorphic as well. In contrast, a real-valued holomorphic function is always constant.

Example 2.8 Let $h = (h_1, \ldots, h_q) : \Omega \to \mathbb{C}^q$ be a holomorphic mapping on an open set Ω in $\mathbb{C}^n$. Let f be a complex-valued C^2-smooth function defined on a neighborhood of the image of h in $\mathbb{C}^q$. Since f is q-holomorphic on $\mathbb{C}^q$ by definition, property 4 in Proposition 2.2 implies that the composition $f \circ h$ is q-holomorphic on Ω. For instance, the Euclidean norm $|h|^2 = \sum_{j=1}^{q} |h_j|^2$ of h is q-holomorphic.

Now fix a positive constant $c > 0$ and define

$$z \mapsto \varphi_c(z) = \frac{1}{1 + c\|h(z)\|^2}.$$

It is q-holomorphic on Ω for the same reasons as above, and if c tends to $+\infty$, then $\{\varphi_c\}_{c>0}$ decreases on Ω to the characteristic function of $A = \{h = 0\}$ given by

$$\chi_A(z) = \begin{cases} 1, & \text{if } z \in A \\ 0, & \text{if } z \in \Omega \setminus A \end{cases}.$$

Example 2.9 If a C^2-smooth function is locally holomorphic with respect to the first $n - q$ variables $z_1, \ldots, z_{n-q}$ of a local chart $z_1, \ldots, z_n$, then it is q-holomorphic.

This example motivated Basener to study the converse statement. Out of the previous examples and the properties of q-holomorphic functions, he concluded that in some cases, q-holomorphic functions contain holomorphicity. More precisely, he showed that a 1-holomorphic function defined on some open set in $\mathbb{C}^2$, whose derivative $\overline{\partial} f$ never vanishes, is holomorphic on leaves of a local foliation by holomorphic curves. Later, Bedford and Kalka spread a series of statements in [19] with conditions on smooth functions satisfying certain non-linear Cauchy–Riemann equations in order to admit a local complex foliation by submanifolds. We only repeat their result involving q-holomorphic functions ([19, Theorem 5.3]).

Theorem 2.15 *Let* $q \in \{1, \ldots, n-1\}$. *Let* f *be a* C^3-*smooth q-holomorphic function on a domain* Ω *in* $\mathbb{C}^n$ *which additionally fulfills*

$$\partial \overline{f} \wedge \overline{\partial} f \wedge (\partial \overline{\partial} f)^{q-1} \neq 0 \quad and \quad (\partial \overline{\partial} f)^q \neq 0.$$

Suppose that $\mathrm{Re}(f)$ *is plurisubharmonic on* Ω. *Then* f *is holomorphic on a foliation by complex submanifolds in* Ω *of codimension q.*

We devote the next chapter to the study of generalized pseudoconvexity and the existence of a complex foliation of complements of such sets.

Chapter 3
Analytic Continuation and
q-Pseudoconvexity

In 1955, Rothstein [150] introduced the notion of q-*convexity* in terms of generalized Hartogs figures (see Definition 3.1) in order to study limits of analytic varieties. Based on this, Grauert [73] was interested in cohomology of complex spaces and used the q-convexity as a boundary condition in the spirit of Levi and in terms of exhaustion functions. Meanwhile, Fujita [68] investigated the continuity principle on *domaines pseudoconvexes d'ordre général* or *d'ordre n − q*. Tadokoro [163] pointed out that these notions are indeed equivalent.

Later, Eastwood and Suria [55] continued Grauert's study on the relation of sheaf cohomology of Rothstein's q-convexity, which they called *Hartogs q-pseudoconvex*. In 1986, Słodkowski [160] defined q-pseudoconvexity in the spirit of Oka under the condition that $- \log \operatorname{dist}(z, \partial\Omega)$ is q-plurisubharmonic. He verified its equivalence to a continuity principle including $(q + 1)$-dimensional analytic sets. Suria [162] proved that, in the case of a smoothly bounded domain in $\mathbb{C}^n$, the notion of Levi q-pseudoconvexity and q-pseudoconvexity by means of the existence of a q-plurisubharmonic exhaustion function are equivalent. Basener [18] produced generalized holomorphically convex hulls of compact sets using q-holomorphic functions, which led to the notion of q-*holomorphic convexity*. It is related to strict q-pseudoconvexity which has been also studied by Hunt and Murray [97] in the context of the Dirichlet problem, peak points and the Shilov boundary. While in the pseudoconvex case $q = 0$ the literature is richly filled by monographs on this topic, there exist—to our knowledge—only a few books which treat q-(pseudo)convexity. These are by Grauert [77], Henkin and Leiterer [90] and Forstnerič [62]. Another survey book is by Colţoiu [42].

In this chapter, we present a list of all the different equivalent notions of q-pseudoconvexity. Detailed proofs and more properties can be found in [139]. Simple examples of q-pseudoconvex domains derive from sublevel sets of q-plurisubharmonic functions or, in the smooth case, defining functions whose Levi matrices have at most q negative eigenvalues. Furthermore, *non-q*-pseudoconvexity admits a duality principle with respect to the $\tilde{q}$-pseudoconvexity of its complement

© The Author(s), under exclusive license to Springer Nature Singapore Pte Ltd. 2022   25
T. Ohsawa, T. Pawlaschyk, *Analytic Continuation and q-Convexity*, SpringerBriefs
in Mathematics, https://doi.org/10.1007/978-981-19-1239-9_3

(with the appropriate index $\tilde{q}$). Roughly speaking, a set is not q-pseudoconvex near its boundary point p if and only if some part of its complement near p touches a strictly $(n-q-2)$-pseudoconvex set from inside (cf. [18, 160]). Using this duality, we discuss the relation of the analytic structure of surfaces (such as graphs of certain continuous mappings) to the q-pseudoconvexity of its complement, i.e., its *q-pseudoconcavity*. The discussion leads to generalizations of Hartogs' theorem and results by Shcherbina [154] and Chirka [38]. They form one of the main results in [133] and [140] and will be presented later in Theorem 3.7 and Corollary 4.2, respectively. As mentioned before, this chapter and the previous one are partially extracted from the doctoral thesis [139] and the papers [141–143].

3.1 q-Pseudoconvex Sets

We introduce the q-pseudoconvexity in the sense of Rothstein [150] which is based on generalized Hartogs figures. Later, it was studied by Eastwood and Suria [55, 162] who named it *Hartogs q-pseudoconvexity*. To distinguish the different notions of q-pseudoconvexity here we use their notation.

Definition 3.1   1. We write $\Delta_r^n := \Delta_r^n(0) = \{z \in \mathbb{C}^n : \max_j |z_j| < r\}$ for the polydisc with radius $r > 0$. We set $A_{r,R}^n := \Delta_R^n \setminus \overline{\Delta_r^n}$.
2. Let $n \geq 2$ and $q \in \{1, \ldots, n-1\}$. Fix two real numbers r, R with $0 < r, R < 1$. A *Euclidean $(n-q, q)$ Hartogs figure* H_e is the set

$$H_e := \left(\Delta_1^{n-q} \times \Delta_r^q\right) \cup \left(A_{R,1}^{n-q} \times \Delta_1^q\right) \subset \Delta_1^{n-q} \times \Delta_1^q = \Delta_1^n.$$

3. A pair (H, P) of domains H and P in $\mathbb{C}^n$ with $H \subset P$ is called a *(general) $(n-q, q)$ Hartogs figure* if there is a Euclidean $(n-q, q)$ Hartogs figure H_e and a biholomorphic mapping F from Δ_1^n onto P such that $F(H_e) = H$.
4. An open set Ω in $\mathbb{C}^n$ is called *Hartogs q-pseudoconvex* if it admits the Kontinuitätssatz with respect to the $(n-q)$-dimensional polydiscs, i.e., given any $(n-q, q)$ Hartogs figure (H, P) such that $H \subset \Omega$, we already have that $P \subset \Omega$.

Another more advanced construction of Hartogs figures but for different purposes is due to Grauert and Riemenschneider [79].

Notice that in regards to the classical Kontinuitätssatz and the solution of the Levi problem, a domain in $\mathbb{C}^n$ is Hartogs $(n-1)$-pseudoconvex if and only if it is pseudoconvex.

Fujita [68] and Tadokoro [163] used a continuity principle derived from the above-mentioned generalized Hartogs figures (cf. Theorem 3.3, property 10).

Using the q-holomorphic functions, Basener [18] introduced the *holomorphically q-convex* sets, i.e., open sets Ω such that with each compact set K in Ω its *q-holomorphically convex hull*

$$\hat{K}_{\Omega,q} = \{z \in \Omega : |f(z)| \leq \max_K |f| \ \forall \ f : \Omega \to \mathbb{C} \ q\text{-holomorphic}\}$$

lies relatively compact in Ω. He compared them to *(strictly) Levi q-peudoconvex sets* (recall Definition 2.2) and clarified the relation of these two notions [18, Theorem 3].

Theorem 3.1 *Let Ω be a bounded domain with C^2-smooth boundary.*

1. *If Ω is q-holomorphically convex, then it is Levi q-pseudoconvex.*
2. *If Ω is strictly (Levi) q-pseudoconvex, then for each $p \in \partial\Omega$ there is a neighborhood U of p such that $\Omega \cap U$ is q-holomorphically convex.*

In this context, it is natural to compare the q-holomorphically convex hull to the *q-plurisubharmonically convex hull*

$$\hat{K}_{\Omega,q}^{\text{psh}} = \{z \in \Omega : \psi(z) \leq \max_K \psi \ \forall \ \psi : \Omega \to [-\infty, \infty) \ q\text{-plurisubharmonic}\}.$$

Since the $\log|f|$ of a given q-holomorphic function f is q-plurisubharmonic, we conclude that the hull $\hat{K}_{\Omega,q}^{\text{psh}}$ is contained in the hull $\hat{K}_{\Omega,q}$, but it is not known to us whether these two hulls coincide. In general, it is up to now an open question whether the converse of the first statement in Theorem 3.1 holds true. Nonetheless, it was shown in [98] that strict q-pseudoconvexity implies $(q + 1)$-holomorphic convexity, and, moreover, that every open set in $\mathbb{C}^n$ is $(n-1)$-holomorpically convex.

Grauert [73] introduced the *q-convexity* in order to study the sheaf cohomology of complex spaces. He calls a domain Ω *q-convex* if it admits a smooth exhaustion function ϱ such that its Levi form has at least $(n - q + 1)$ positive eigenvalues outside a compact set K. If K is empty, Ω is called *q-complete*. Thus, the $(q + 1)$-completeness is the same as demanding the existence of a smooth strictly q-plurisubharmonic exhaustion function in our notation.

Suria's intention in [162] was to clarify the relation of Levi q-pseudoconvexity and $(q + 1)$-completeness.[1] Together with results from [55] we have the following statement.

Theorem 3.2 *Let Ω be a C^2-smoothly bounded domain in $\mathbb{C}^n$. Then Ω is Levi q-pseudoconvex if and only if it admits a C^2-smooth q-plurisubharmonic exhaustion function.*

In his proof, Suria carefully constructs a smooth q-plurisubharmonic exhaustion function out of the Levi q-pseudoconvex boundary data. In the strict case, we obtain more. First, assume that the domain Ω is strictly q-pseudoconvex at a boundary point p and let ψ be a local defining function for Ω at p defined on a neighborhood

---

[1] "This seems to be one of those facts that every complex analyst believes, perhaps for psychological reasons, but no precise reference is, to my knowledge, available and all mathematicians whom I have asked so far don't seem to know how a precise proof should go." (Suria, [162, p. 105])

U of p. Then for a large enough constant $c > 0$ the function $\exp(c\psi) - 1$ is strictly q-plurisubharmonic on some ball $B \Subset U$ centered at p and still locally defines Ω at p. Second, if the domain Ω is bounded and strictly q-pseudoconvex at each of its boundary points, it is now possible to glue together the local strictly q-plurisubharmonic defining functions from the previous discussion to a *global defining function*, i.e., there exists a C^2-smooth function ϱ defined on a neighborhood V of the closure $\overline{\Omega}$ such that $\Omega = \{z \in V : \varrho(z) < 0\}$, $d\varrho \neq 0$ on $\partial\Omega$, and ϱ is strictly q-plurisubharmonic on V. This procedure is well-known in the pseudoconvex case $q = 0$ (cf. Proposition 3.2.2 in [112]) and transfers easily to the case $q \geq 1$. But if the domain Ω is *unbounded*, the existence of a global defining function is no longer guaranteed. Indeed, the domain $\Omega = \{(z, w) \in \mathbb{C}^2 : |w| < e^{-|z|^2}\}$ is strictly pseudoconvex and contains a copy $L = \mathbb{C} \times \{0\}$ of $\mathbb{C}$. This implies that each bounded above, strictly plurisubharmonic function ψ on Ω has to be constant on L. But this contradicts the strict plurisubharmonicity of ψ. Nevertheless, if Ω is strictly q-pseudoconvex, there exists a global q-plurisubharmonic defining function on a neighborhood of $\overline{\Omega}$ which is strictly q-plurisubharmonic near $\partial\Omega$. The proof is due to Harz, Shcherbina and Tomassini [89].

Now that we have recalled many notions of generalized pseudoconvexity, it is time to clarify, which ones are equivalent. Detailed proofs or partial results can be found in, e.g., [53, 103, 141, 142, 160] or altogether collected in [139].

Theorem 3.3 *Let $q \in \{0 \ldots, n-2\}$ and Ω be an open set in $\mathbb{C}^n$. Then the following statements are all equivalent:*

1. *The set Ω is Hartogs $(n-q-1)$-pseudoconvex.*
2. *For every vector X in $\mathbb{C}^n$ with Euclidean norm $|X| = 1$, the $-\log$ of the Hartogs radius function in X-direction $z \mapsto -\log \mathrm{dist}_X(z, \partial\Omega)$ is q-plurisubharmonic on Ω. Here, $\mathrm{dist}_X(z, \partial\Omega) = \mathrm{dist}\left(z, \partial\Omega \cap (z + \mathbb{C}X)\right)$.*
3. *For every complex norm $\|\cdot\|$ the function*

$$z \mapsto -\log d_{\|\cdot\|}(z, \partial\Omega) = -\log \inf\{\|z - w\| : w \in \partial\Omega\}$$

   is q-plurisubharmonic on Ω.
4. *For some complex norm $\|\cdot\|$ the function $-\log d_{\|\cdot\|}(z, \partial\Omega)$ is q-plurisubharmonic on Ω.*
5. *The function $-\log \mathrm{dist}(z, \partial\Omega)$ is q-plurisubharmonic on Ω, where $\mathrm{dist}(z, \partial\Omega)$ is the Euclidean distance from z to the boundary of Ω.*
6. *Ω admits a continuous q-plurisubharmonic exhaustion function.*
7. *There exists a (not necessarily continuous) q-plurisubharmonic function ψ on Ω such that ψ tends to $+\infty$ whenever z approaches $\partial\Omega$.*
8. *There exist a neighborhood W of $\partial\Omega$ and a q-plurisubharmonic function ψ on $W \cap \Omega$ such that ψ tends to $+\infty$ whenever z approaches $\partial\Omega$.*
9. *For every compact set K in Ω, its q-plurisubharmonic hull $\hat{K}_{\Omega,q}^{psh}$ is compact.*
10. *(Continuity principle) Let $\{A_t\}_{t \in [0,1]}$ be a family of analytic subsets in some open set U in $\mathbb{C}^n$ which continuously depend on t in the Hausdorff topology.*

Assume further that $\dim_z A_t \geq q + 1$ *for every* $z \in A_t$ *and* $t \in [0, 1]$, *and that the closure of* $\bigcup_{t \in [0,1]} A_t$ *is compact. If* Ω *contains the boundary* ∂A_1 *and the closure* $\overline{A_t}$ *for each* $t \in [0, 1)$, *then the closure* $\overline{A_1}$ *also lies in* Ω.

We fix what we mean by *q*-pseudoconvexity.

Definition 3.2 An open set Ω in $\mathbb{C}^n$ is said to be *q-pseudoconvex* (in the sense of Słodkowski, cf. [160] and in the spirit of Oka), if it fulfills property 5 of Theorem 3.3.

Using Theorem 3.3, we obtain more equivalent properties.

Proposition 3.1 *Let* $q \in \{0 \ldots, n - 2\}$ *and* Ω *be an open set in* $\mathbb{C}^n$. *Then* Ω *is q-pseudoconvex if one of the following statements holds true:*

1. *For every point* p *in* $\partial\Omega$ *there is a ball* $B = B_r(p)$ *centered at* p *such that* $\Omega \cap B$ *is q-pseudoconvex.*
2. *There is a collection* $\{\Omega_j\}_{j \in \mathbb{N}}$ *of bounded q-pseudoconvex domains* Ω_j *in* Ω *such that* $\Omega_j \Subset \Omega_{j+1} \Subset \Omega$ *for every* $j \in \mathbb{N}$ *and* $\Omega = \bigcup_{j \in \mathbb{N}} \Omega_j$.
3. *The intersection* $\Omega \cap \pi$ *is q-pseudoconvex in* π *for every* $(q + 2)$-*dimensional complex affine subspace* π *in* $\mathbb{C}^n$.

We can improve the regularity of the exhaustion functions of a *q*-pseudoconvex set. By Bungart's approximation technique (see Theorem 2.5) the exhaustion function of a *q*-pseudoconvex set can be assumed to be piecewise smooth. In the case of $q = 0$, due to Richberg's approximation for plurisubharmonic functions, we can even obtain that an open set Ω is pseudoconvex if and only if it has a C^∞-smooth plurisubharmonic exhaustion function. If $q \geq 1$, it is unknown whether a *q*-pseudoconvex set admits a C^1-smooth *q*-plurisubharmonic exhaustion function. Due to the discussion after Theorem 3.1 on the relation of the hulls created by *q*-holomorphic and *q*-plurisubharmonic functions, it follows from Theorem 3.3, point 9, that a *q*-holomorphically convex set (with not necessarily smooth boundary) is *q*-pseudoconvex.

Properties of *q*-plurisubharmonic functions and Theorem 3.3 now imply the following Behnke–Sommer type statements.

Proposition 3.2

1. *If* Ω_1 *is q-pseudoconvex and* Ω_2 *is r-pseudoconvex in* $\mathbb{C}^n$, *then the union* $\Omega_1 \cup \Omega_2$ *is* $(q + r + 1)$-*pseudoconvex.*
2. *Let* $\{\Omega_j\}_{j \in J}$ *be a collection of q-pseudoconvex sets in* $\mathbb{C}^n$ *such that the interior* Ω *of the intersection* $\bigcap_{j \in J} \Omega_j$ *is not empty. Then* Ω *is q-pseudoconvex.*
3. *If* $\{\Omega_j\}_{j \in \mathbb{N}}$ *is an increasing collection of q-pseudoconvex sets in* $\mathbb{C}^n$ *with* $\Omega_j \subset \Omega_{j+1}$ *for each* j, *then the union* $\Omega = \bigcup_{j \in \mathbb{N}} \Omega_j$ *is q-pseudoconvex.*

Since we will examine subdomains which are *q*-pseudoconvex only in a neighborhood of particular points of their boundary relative to a prescribed larger domain, we recall the notion of *relative q-pseudoconvexity* introduced by Słodkowski [160].

Definition 3.3 Let $U \subsetneq V$ be two open sets in $\mathbb{C}^n$. We say that U is q-*pseudoconvex in V* if for every point p in $V \cap \partial U$ there exists an open ball $B_r(p)$ centered in p such that the intersection $U \cap B_r(p)$ is q-pseudoconvex according to Definition 3.2.

The next characterization of relative q-pseudoconvex sets is similar to property 8 of Theorem 3.3 and serves as useful tool in the forthcoming chapter.

Proposition 3.3 Let $U \subsetneq V$ be open sets in $\mathbb{C}^n$. Then the following statements are equivalent:

1. U is q-pseudoconvex in V.
2. There exist a neighborhood W of $\partial U \cap V$ in V and a q-plurisubharmonic function ψ on $W \cap U$ such that $\psi(z)$ tends to $+\infty$ whenever z approaches the relative boundary $\partial U \cap V$.

The q-pseudoconvex sets in $V = \mathbb{C}^n$ are simply the q-pseudoconvex sets defined in the beginning of this chapter.

Example 3.1 An easy example quickly derives from sublevel sets of q-plurisubharmonic functions. Let φ be q-plurisubharmonic on an open set V in $\mathbb{C}^n$ and let c be a real number. Then, the set $U = \{z \in V : \varphi(z) < c\}$ is q-pseudoconvex in V. If, moreover, the set V is itself q-pseudoconvex, then U is also q-pseudoconvex (in $\mathbb{C}^n$).

A more interesting example reads as follows.

Example 3.2 Let Ω be an open set in $\mathbb{C}^n$ and let h be a smooth q-holomorphic function on Ω (recall our Sect. 2.4). Let $G_h = \{(z, h(z)) \in \mathbb{C}^{n+1} : z \in \Omega\}$ be the graph of h over Ω. Then the function $(z, w) \mapsto 1/(h(z) - w)$ is q-holomorphic on $U := (\Omega \times \mathbb{C}) \setminus G_h$, so the function $\psi(z, w) := -\log|h(z) - w|$ is q-plurisubharmonic on U. It has the property that $\psi(z, w)$ tends to $+\infty$ whenever (z, w) approaches the graph G_h. Hence, the open set U is q-pseudoconvex in $V := \Omega \times \mathbb{C} \subset \mathbb{C}^{n+1}$ by Proposition 3.3. A converse statement is not known except for Hartogs' theorem, which appears in the holomorphic case $q = 0$. However, if the complement of the graph G_h is q-pseudoconvex in V, then in view of point 2 in Theorem 3.3, we have that

$$\varphi(z) = -\log \operatorname{dist}_X(z, \partial\Omega) = -\log|h(z)|$$

is q-plurisubharmonic on Ω for the direction $X = (0, 1) \in \mathbb{C}_z^n \times \mathbb{C}_w$. If we additionally assume that G_h^c is $(n - q - 1)$-pseudoconvex, then by the same reasoning, φ is $(n - q - 1)$-plurisubharmonic. Therefore, it fulfills $(\partial\bar\partial\varphi)^n = 0$, but it is not clear if these are sufficient conditions on the existence of a foliation of G_h by complex submanifolds.

Again, we refer to [19] for more conditions on foliations of graphs in the context of q-holomorphic functions. We discuss this problem for q-pseudoconvex complements of graphs in the next section.

3.2 q-Pseudoconcave Sets and Foliations

Complements of q-pseudoconvex sets have been already studied by Basener [18] and Słodkowski [160]. In order to obtain properties of the Shilov boundary of product sets, Słodkowski introduced the notion of k-*maximum sets*, i.e., closed subsets X of an open set V in $\mathbb{C}^n$ such that for every complex k-codimensional affine subspace π the complex polynomials admit the local maximum modulus property on the intersection $X \cap \pi$. An instance of a k-maximum set is given by analytic subsets $X = A$ of V such that $\dim_z A \geq k + 1$ for each $z \in A$. In fact, k-maximum sets are closely related to q-plurisubharmonicity and q-pseudoconvexity.

Theorem 3.4 *Fix $k \in \{0, \ldots, n-1\}$. Let $X = V \cap \overline{X}$ be a locally closed set in $\mathbb{C}^n$. Then X is k-maximum if and only if it admits one of the following properties:*

1. *The k-plurisubharmonic functions on V admit the local maximum property on X (cf. Theorem 2.11).*
2. *The characteristic function χ_X is $(n-k-1)$-plurisubharmonic on V. Recall that*

$$\chi_X(z) = \begin{cases} 0, & z \in X \\ -\infty, & z \in V \setminus X \end{cases}.$$

3. *The set $U = V \setminus X$ is $(n-k-2)$-pseudoconvex in V.*

Let us denote only here by $\Pi := \hat{K}^{\text{psh}}_{\mathbb{C}^n, n-q-2}$ the $(n-q-2)$-plurisubharmonic hull of a compact set K in $\mathbb{C}^n$. In [142, Corollary 5.6], it was shown that $X = \Pi \setminus K$ is $(n-q-2)$-maximum. By the previous theorem, $\mathbb{C}^n \setminus X$ is q-pseudoconvex in $\mathbb{C}^n \setminus K$. From this, it follows that, if Ω is a q-pseudoconvex domain in $\mathbb{C}^n$ such that K lies in the complement of Ω, then the set $\Omega \setminus \Pi$ is again q-pseudoconvex. This result was already known in the case $q = 0$ and $n = 2$ (cf. [158, Theorem 2.1]).

Słodkowski [160] showed that q-pseudoconvexity is also equivalent to a *weak* and a *strong* version of the Kontinuitätssatz. The strong one is similar to Hartogs $(n-q-1)$-pseudoconvexity, while the weak one resembles the continuity principle with analytic varieties of dimension $q+1$ (cf. Theorem 3.3, point 10). Earlier, Basener [18] discovered that the continuity principle appears in the complement of a strictly q-pseudoconvex sets. The main idea is that, locally, the boundary of a strictly q-pseudoconvex set is the graph of a real-valued function with a certain number of positive eigenvalues. This observation leads to a duality principle between q-pseudoconvex sets and their complements.

Proposition 3.4 *If an open set Ω in $\mathbb{C}^n$ is strictly q-pseudoconvex at some point $p \in \partial\Omega$, then for every small enough neighborhood V of p, for each w in $\partial\Omega \cap V$ and every neighborhood $W \Subset V$ of w there is a family $\{A_t\}_{t \in [0,1]}$ of $(n-q-1)$-dimensional complex submanifolds of W which is continuously parameterized by t and fulfills:*

- $\overline{A}_t \subset (\mathbb{C}^n \setminus \overline{\Omega})$ *for every $t \in [0, 1)$;*
- *and $w \in A_1$, but $\overline{A}_1 \setminus \{w\} \subset (\mathbb{C}^n \setminus \overline{\Omega})$.*

In view of Theorem 3.3, this means that $U = V \setminus \overline{\Omega}$ is not $(n-q-2)$-pseudoconvex in V.

It was verified in [139, 140] that a converse version of this duality principle holds true. It is obtained by a careful investigation of the generalized Hartogs figures near boundary points at which the given domain is *not* q-pseudoconvex.

Theorem 3.5 *Let Ω be a domain in $\mathbb{C}^n$ which is not q-pseudoconvex. Then there exist a point $p \in \partial\Omega$, a neighborhood V of p and a strictly $(n-q-2)$-pseudoconvex set G in V such that the set $V \setminus \Omega$ is contained in $G \cup \{p\}$ and $\{p\} = \partial G \cap \partial\Omega \cap V$, i.e., $V \setminus \Omega$ touches ∂G from the inside of G only at p.*

Observe that in this situation, according to the discussion after Theorem 3.2, there exists a smooth strictly $(n-q-2)$-plurisubharmonic function near p which *peaks at p on G*, i.e., $\psi(p) = 0$ and $\psi < 0$ near p on G.

The next part is an outline of the main result in [140] on the complex structure of surfaces of higher codimension in $\mathbb{C}^n$. It deals with the following issue: Let $U \subset V$ be two domains in $\mathbb{C}^n$, $n \geq 2$, and let U be q-pseudoconvex in V. Does it follow that the set $G = V \setminus U$ possesses an analytic structure? Here, by the existence of an *analytic structure* on a locally closed set $G \subset \mathbb{C}^n$ we mean that for every point $z \in G$ there is a complex analytic variety $A_z \subset G$ of positive dimension containing z. First, we clarify the converse situation.

Proposition 3.5 *Let $q \in \{1, \ldots, n-1\}$ and let S be a closed subset of an open set Ω in $\mathbb{C}^n$. Assume that the boundary $\partial_\Omega S$ of S in Ω is locally filled by analytic sets of dimension at least q, i.e., for every point p in $\partial_\Omega S$ there is a neighborhood W of p in Ω such that for each point w in $\partial_\Omega S \cap W$ there exists an analytic subset A_w of W with $w \in A_w \subset S$ and $\dim_z A_w \geq q$ for each $z \in A_w$. Then $\Omega \setminus S$ is k-pseudoconvex in Ω with $k = n-q-1$.*

Indeed, if not, then by the duality principle in Theorem 3.5 and its subsequent remark, there exist an analytic set A in S of dimension $\geq q = n-k-1$ and an $(n-k-2)$-plurisubharmonic function ψ near some point $p \in A$ which peaks at p in A. But this violates the local maximum principle on analytic sets (cf. Theorem 2.11).

We have already encountered special cases of the initial question: Hartogs' theorem (Theorem 1.3), Theorem 1.11 and Theorem 2.15. A similar statement to that of Hartogs holds true for graphs of real codimension one. The following result was first proved in a fundamental paper of Levi [116] from 1911 for smooth graphs in the special case $n = 1$. It was established much later by Shcherbina in [154] for continuous f in the case $n = 1$. Then, using the latter one, Chirka [38] generalized it to the case $n > 1$.

Theorem 3.6 *Let $f : B' \to \mathbb{R}_v$ be a continuous function defined on the unit ball B' in $\mathbb{C}_z^n \times \mathbb{R}_u$. Then the complement of its graph*

$$G_f = \{(z, w) : (z, u) \in B', \ v = f(z, u), \ w = u + iv\}$$

in $(B' \times \mathbb{R}_v) \subset \mathbb{C}_z^n \times \mathbb{C}_w$ is a disjoint union of two domains of holomorphy if and only if the graph G_f can be decomposed into a disjoint union of complex analytic hypersurfaces.

More details on the theorems of Hartogs and Levi and their complete proofs can be found in [153, Theorem 2 on p. 226] and [166, Levi's theorem on p. 164], respectively.

In order to simplify our notations, we introduce a generalized version of concavity.

Definition 3.4 Let $q \in \{0, \dots, n-1\}$ and let S be a closed subset of an open set V in $\mathbb{C}^n$. We say that S is a *Hartogs q-pseudoconcave subset of V* if $U := V \setminus S$ is $(n-q-1)$-pseudoconvex in V.

Let us first treat the case of smooth surfaces. Freeman [63] gave necessary and sufficient conditions on a real C^2-smooth submanifold $\Gamma = \{\varphi_1 = \dots = \varphi_r = 0\}$ in $\mathbb{C}^n$ to admit a local complex foliation. For instance, if Γ is a CR manifold of CR dimension q, and if the holomorphic tangent space

$$H_p \Gamma := \bigcap_{j=1}^{r} \left\{ Z \in \mathbb{C}^n : \sum_{k=1}^{n} \frac{\partial \varphi_j}{\partial z_k}(p) Z_k = 0 \right\}$$

and the *Levi null space*

$$N_p := \bigcap_{j=1}^{r} \left\{ X \in H_p \Gamma : \mathcal{L}_{\varphi_j}(p)(X, Y) = 0 \text{ for every } Y \in H_p \Gamma \right\}$$

coincide, there exists a local complex foliation by q-dimensional submanifolds (cf. Theorem 1.1 in [63]). In fact, the condition on the Levi null space can be replaced by assuming that Γ is Hartogs q-pseudoconcave with the same outcome. Indeed, if we assume that in this case the Levi null space is a proper subset of $H_p \Gamma$, we can construct a strictly $(q-1)$-pseudoconvex *tubular* neighborhood of Γ so that Γ touches it from inside at p (see [140] or Proposition 4.7.5 in [139] for a detailed proof). By applying the duality principle in Proposition 3.4, we get a contradiction. Hence, $H_p = N_p$, so there is the desired foliation due to Freeman.

Now we turn to the continuous case. Locally, a generic smooth submanifold of codimension k in $\mathbb{C}^N$ can be represented by the graph of a smooth mapping $f : U \to \mathbb{R}^k$, where U is an open set in $\mathbb{C}^n \times \mathbb{R}^k$ with $n = N - k$ (c.f., e.g. [8, §1.3]). This representation is not unique in general, but it motivates us to define non-smooth surfaces locally as graphs of continuous mappings $f : U \to \mathbb{R}^k$. For this reason, our main setting reads as follows:

Fix integers $n \geq 1$ and $k, p \geq 0$ such that $N = n + k + p \geq 2$. Then $\mathbb{C}^N$ splits into the product

$$\mathbb{C}^N = \mathbb{C}^n_z \times \mathbb{C}^k_{w=u+iv} \times \mathbb{C}^p_\zeta = \mathbb{C}^n_z \times (\mathbb{R}^k_u + i\mathbb{R}^k_v) \times \mathbb{C}^p_\zeta,$$

where $w = u + iv$. Let D be an open set in $\mathbb{C}^n_z \times \mathbb{R}^k_u$ and let $f = (f_v, f_\zeta)$ be continuous on D with image in $\mathbb{R}^k_v \times \mathbb{C}^p_\zeta$. Then the graph of f is given by

$$G_f = \{(z, w, \zeta) \in \mathbb{C}^n_z \times \mathbb{C}^k_w \times \mathbb{C}^p_\zeta : (z, u) \in D, \ (v, \zeta) = f(z, u)\}.$$

Moreover, we denote by $\pi_{z,u}$ the natural projection

$$\pi_{z,u} : \mathbb{C}^n_z \times \mathbb{C}^k_w \to \mathbb{C}^n_z \times \mathbb{R}^k_u, \quad \pi_{z,u}(z, w) \mapsto (z, u).$$

We are interested in the question whether G_f admits a local foliation by complex submanifolds. In this context, we have first to study the Hartogs q-pseudoconcavity of the graph of f. Notice that we prefer the notion of Hartogs q-pseudoconcavity since it fits better to the dimension of the desired foliation.

Lemma 3.1 *Let N, n, k, p, D be from above. Pick another integers $m \in \{1, \ldots, n\}$ and $r \in \{0, \ldots, p\}$ with $k + r \geq 1$. For $\mu_1, \ldots, \mu_r \in \{1, \ldots, p\}$ with $\mu_1 < \ldots < \mu_r$, we divide the coordinates of ζ into $\zeta' = (\zeta_{\mu_1}, \ldots, \zeta_{\mu_r})$ and the remaining coordinates $\zeta'' = (\zeta_j : j \in \{1, \ldots p\} \setminus \{\mu_1, \ldots, \mu_r\})$ which we assume to be ordered by their index j, as well. Finally, let Π be a complex m-dimensional affine subspace in $\mathbb{C}^n_z$. We set $D_\bullet := D \cap (\Pi \times \mathbb{R}^k_u)$ and $f_\bullet := (f_v, f_{\zeta'})|_{D_\bullet}$. If the graph G_f is Hartogs n-pseudoconcave in $D \times \mathbb{R}^k_v \times \mathbb{C}^p_\zeta$, then the graph $G_{f_\bullet}$ is Hartogs m-pseudoconcave in $D_\bullet \times \mathbb{R}^k_v \times \mathbb{C}^r_{\zeta'}$.*

Proof Since the Hartogs n-pseudoconcavity is a local property, after shrinking D if necessary and after a biholomorphic change of coordinates we can assume without loss of generality that $\Pi = \{0\}^{n-m}_{z'} \times \mathbb{C}^m_{z''} \subset \mathbb{C}^n_z$, where $z' = (z_1, \ldots, z_{n-m})$ and $z'' = (z_{n-m+1}, \ldots, z_n)$, and that the ζ-coordinates are ordered in such a way that $\zeta' = (\zeta_1, \ldots, \zeta_r)$ and $\zeta'' = (\zeta_{r+1}, \ldots, \zeta_p)$.

Assume that $G_{f_\bullet}$ is not Hartogs m-pseudoconcave in $D_\bullet \times \mathbb{R}^k_v \times \mathbb{C}^r_{\zeta'}$, and let $M := m + k + r$. Then in view of Theorem 3.3, there are a point p in $G_{f_\bullet}$ and a ball $B = B_\varrho(p)$ in $\mathbb{C}^M$ such that the set $(\mathbb{C}^M \setminus G_{f_\bullet}) \cap B$ is not $(M - m - 1) = (k + r - 1)$-pseudoconvex. Since B is pseudoconvex, according to the continuity principle in Theorem 3.3 there is a family $\{A_t\}_{t \in [0,1]}$ of analytic sets A_t in $\mathbb{C}^M$ which depends continuously on t and which fulfills the following properties:

- $\dim_z A_t \geq k + r$ for every $z \in A_t$ and $t \in [0, 1]$;
- the closure of the union $\bigcup_{t\in[0,1]} A_t$ is compact;
- for every $t \in [0, 1)$ the intersection $\overline{A_t} \cap G_{f_\bullet}$ is empty;
- $\partial A_1 \cap G_{f_\bullet}$ is empty, as well;
- and the set A_1 touches $G_{f_\bullet}$ at a point $p_0 = (z_0, w_0, \zeta_0')$, where $z_0 = (z_0', z_0'') = (0, z_0'')$ and $w_0 = u_0 + i v_0$.

Given some positive number $a > 0$, consider the analytic sets

$$S_t := \{0\}^{n-m} \times A_t \times \Delta_a^{p-r}(f_{\zeta''}(z_0, u_0)) \subset \mathbb{C}^N.$$

It is easy to verify that each S_t is an analytic set of dimension at least $k + p$ and that the family $\{S_t\}_{t\in[0,1]}$ violates the continuity principle in Theorem 3.3 so that the complement of G_f cannot be $(k+p-1)$-pseudoconvex in $D \times \mathbb{R}_v^k \times \mathbb{C}_\zeta^p$. Therefore, G_f cannot be Hartogs $n = (N-(k+p-1)-1)$-pseudoconcave, which is a contradiction to the assumption made on G_f. This means that $G_{f_\bullet}$ has to be Hartogs m-pseudoconcave. □

We are now able to state the main theorem in [140] and give a sketch of its proof. In the next chapter, we discuss a similar result (cf. Corollary 4.2) using completely different methods from q-convexity in the sense of Grauert and L^2-theory.

Theorem 3.7 *Let n, k, p be integers with $n \geq 1$, $p \geq 0$ and $k \in \{0, 1\}$ such that $N = n + k + p \geq 2$. Let D be a domain in $\mathbb{C}_z^n \times \mathbb{R}_u^k$ and let $f : D \to \mathbb{R}_v^k \times \mathbb{C}_\zeta^p$ be a continuous function such that G_f is Hartogs n-pseudoconcave. Then G_f is locally the disjoint union of n-dimensional complex submanifolds.*

Sketch of Proof The case **n** $\geq$ **1, k** = **0, p** = **1** is the classical Hartogs' theorem.

For the case **n** $\geq$ **1, k** = **0, p** $\geq$ **1**, apply Lemma 3.1 on each component function $f_j := f_{\zeta_j}, j = 1, \ldots, p$, in order to obtain that for each j the complement of the graph of f_j is pseudoconvex. By Hartogs' theorem, the f_j's are holomorphic, which means that G_f is a complex surface.

The case **n** = **1, k** = **1, p** = **0** is due to Shcherbina [154].

The case **n** $\geq$ **2, k** = **1, p** = **0** has been proved by Chirka [38].

The proof of the case **n** = **1, k** = **1, p** = **1** is nothing but trivial and involves various techniques from the previous chapters on q-plurisubharmonicity and q-pseudoconvexity. The main idea is first to extract holomorphic curves γ_α from the Hartogs 1-pseudoconcave graph G_{f_v} using Shcherbina's result in the case $n = k = 1$ and $p = 0$. Since these curves are given by graphs $\Gamma(g_\alpha) = \gamma_\alpha$ of holomorphic functions $g_\alpha = u_\alpha + i v_\alpha$, the curves foliating the graph of f can be defined via

$$z \mapsto \big(f_v(z, u_\alpha(z)), f_\zeta(z, u_\alpha(z))\big) = \big(v_\alpha(z), f_\zeta(z, u_\alpha(z))\big) \in \mathbb{R}_v \times \mathbb{C}_\zeta.$$

Now if one of the functions $z \to f_\zeta(z, u_\alpha(z))$ is not holomorphic, by the duality principle in Theorem 3.5 we can construct a strictly 1-pseudoconvex set U such that G_f touches it from inside. But this contradicts the Hartogs 1-pseudoconcavity of G_f.

The case $\mathbf{n \geq 1}$, $\mathbf{k = 1}$, $\mathbf{p = 1}$ uses an approach similar to the previous case. Chirka's result implies that the Hartogs n-pseudoconcave graph G_{f_v} is foliated by a family of complex hypersurfaces A_α. By Hartogs' theorem of separate holomorphicity, the problem is then reduced to the previous case $n = k = p = 1$.

For the case $\mathbf{n \geq 1}$, $\mathbf{k = 1}$, $\mathbf{p \geq 1}$ we simply apply the preceding case $n \geq 1$ and $k = p = 1$ to the graphs $G_{f_v, f_{\zeta_j}}$ for each $j = 1, \ldots, p$.

Now that we have covered every possible case, the sketch of the proof is finished.

$\square$

So far, we do not have techniques to treat the case $n = 1$, $k = 2$ and $p \geq 0$.

Example 3.3 We close this chapter with an example extracted from [99] which demonstrates that it is not always possible to foliate a 1-pseudoconcave real 4-dimensional submanifold in $\mathbb{C}^3$ by complex submanifolds, but it is still possible to do this by analytic subsets.[2] For a fixed integer $k \geq 2$ consider the function

$$f(z, w) := \begin{cases} \bar{z} w^{2+k} / \bar{w}, & w \neq 0 \\ 0, & w = 0 \end{cases}.$$

It is C^k-smooth on $\mathbb{C}^2$ and holomorphic on complex lines passing through the origin, since $f(\lambda v) = \lambda^{2+k} f(v)$ for every $\lambda \in \mathbb{C}^* := \mathbb{C} \setminus \{0\}$ and each vector $v \in \mathbb{C}^2$. Therefore, in view of Sect. 2.4, the function f is 1-holomorphic on $\mathbb{C}^2$, so $\psi(z, w, \zeta) := -\log |f(z, w) - \zeta|$ is 1-plurisubharmonic outside $\{f = \zeta\}$. Due to Theorem 3.3, this means that the graph G_f of f is a Hartogs 1-pseudoconcave real 4-dimensional submanifold of $\mathbb{C}^3$ which does not admit a regular foliation near the origin, but admits a singular one which is given by the family of holomorphic curves $\{G_{f|\mathbb{C}^* v} : v \in \mathbb{C}^2\}$. Of course, the problem arises because the complex Jacobian of f has non-constant rank near the origin.

---

[2] Thanks to Prof. Kang-Tae Kim for attracting our attention to this example.

Chapter 4
q-Convexity and q-Cycle Spaces

As already mentioned previously, the notion of q-*convexity* was developed by Rothstein [150] and Grauert [73]. It was transfered to q-convex spaces by Andreotti and Grauert (1962). Andreotti–Grauert's finiteness theorem was applied by Andreotti and Norguet (1966–1971) to extend Grauert's solution of the Levi problem to q-convex spaces. A consequence is that the sets of $(q-1)$-cycles of q-convex domains with smooth boundary in projective algebraic manifolds, which are equipped with complex structures as open subsets of Chow varieties, are in fact holomorphically convex.

In this chapter, we study complements of analytic curves and explain the relation of q-convexity and cycle spaces, present results for q-convex domains in projective spaces and investigate the q-convexity in analytic families. As an application we derive a Hartogs type result in Corollary 4.2.

4.1 q-Convex Functions and Sets

Rothstein [150] and Grauert [73] introduced the notion of q-convexity in order to generalize the theory of continuation of analytic functions to, respectively, that of analytic sets and analytic cohomology classes.

Definition 4.1 Let $q \in \{1, \ldots, n\}$ and let M be an n-dimensional complex manifold. A C^2-smooth function φ on M is said to be *(weakly) q-convex at a point* $x \in M$ if $\partial\bar{\partial}\varphi$, or the Levi form of φ, has at least $n - q + 1$ positive (resp., nonnegative) eigenvalues at x.

M is called *q-convex* (resp., *q-complete*) if it admits an exhaustion function which is q-convex outside a compact subset of M (resp., everywhere).

Nakano [122] called M *weakly 1-complete* if M admits a C^∞-smooth plurisubharmonic exhaustion function. In accordance with Rothstein, Grauert and Nakano,

© The Author(s), under exclusive license to Springer Nature Singapore Pte Ltd. 2022
T. Ohsawa, T. Pawlaschyk, *Analytic Continuation and q-Convexity*, SpringerBriefs
in Mathematics, https://doi.org/10.1007/978-981-19-1239-9_4

we shall call M *weakly q-complete* if M admits a C^∞-smooth exhaustion function whose Levi form has at most $q - 1$ negative eigenvalues everywhere. Since the composite of a C^2-smooth convex increasing function with a q-convex function is q-convex, q-convex manifolds are weakly q-complete. *Weakly q-convex manifolds* are defined according to Definition 4.1 by weakening the notion of q-convexity, but there is actually no difference between weakly q-complete manifolds and weakly q-convex manifolds. A weakly q-complete manifold is often referred to as (M, φ) with an exhaustion function φ.

These definitions naturally extend to complex spaces if one defines the q-convexity of a function φ on a complex space X by the property that φ extends to a q-convex function on a neighborhood of the image of any local holomorphic embedding of X into a domain of $\mathbb{C}^N$ (compare to Definition 2.4).

Similar to the case of the notion of pseudoconvexity, there are several definitions of q-convexity. O. Fujita [68] and Tadokoro [163] introduced *pseudoconvex sets of order $n - q$* by extending the works of Oka [134, 138] and Nishino [125]. Fujita showed that the domains of existence of normal families of analytic sets of dimension $n - q$ have this property and that this notion is equivalent to the q-convexity of Rothstein and Grauert for the domains in $\mathbb{C}^n$. As an analogue of Oka's theorem, Basener [18] proved a local existence theorem on smoothly bounded q-convex domains for the solutions of the equation $\overline{\partial} f \wedge (\partial \overline{\partial} f)^q = 0$ (cf. Sect. 2.4).

By virtue of the discussions in Chaps. 2 and 3, a smooth $(q-1)$-plurisubharmonic function in the sense of Hunt and Murray [97] is exactly the same as a weakly q-convex function in the setting of $\mathbb{C}^n$. Therefore, every weakly q-complete domain in $\mathbb{C}^n$ is $(q - 1)$-pseudoconvex in the sense of Definition 3.2.

4.2   Andreotti–Grauert Theory and Cycle Spaces

Originally, q-convex spaces and q-complete spaces were introduced by Andreotti and Grauert [2] to generalize the important finiteness theorems of Cartan and Serre [33] and Grauert [72]. Recall that the former is the basis of the Hirzebruch–Riemann–Roch formula (cf. [91]) and the latter gives a solution to the Levi problem on complex spaces which implies, in particular, a geometric characterization of non-singular models of the germs of analytic spaces with isolated singularities (cf. [74]).

Theorem 4.1 (Andreotti–Grauert Finiteness Theorem) *For any coherent analytic sheaf $\mathcal{F}$ over a q-convex space X, the dimension of the k-th cohomology group $H^k(X, \mathcal{F})$ is finite for all $k \geq q$, and $H^k(X, \mathcal{F}) = 0$ for all $k \geq q$ if moreover X is q-complete.*

This was applied by Andreotti and Norguet [3–5] to extend Grauert's solution of the Levi problem to q-convex spaces. A remarkable consequence is that the sets of $(q - 1)$-*cycles* (see below for the definition) of q-convex domains with C^2-smooth boundary in projective algebraic manifolds, which are equipped with complex structures as open subsets of Chow varieties,[1] are *holomorphically convex*.[2] In [3] an observation was made that the complement of a q-codimensional hyperplane in $\mathbb{CP}^n$ is q-complete.

The holomorphic convexity of cycle spaces was further generalized by Barlet [9, 10]. Namely, given any reduced complex space X and any nonnegative integer k, Barlet defined in [9] a structure of a reduced complex space on the set $\mathscr{C}_k(X)$ of formal finite linear combinations $\sum m_j Z_j$ of reduced and irreducible compact complex subspaces Z_j of dimension k with positive integral coefficients m_j. The complex structure of $\mathscr{C}_k(X)$ is roughly described as follows. For $k = 0$ it is defined in an obvious way. For $k \geq 1$, $\mathscr{C}_k(X)$ is defined in such a way that, for every $Z^0 = \sum_j m_j Z_j^0 \in \mathscr{C}_k(X)$, for every $x \in \bigcup_j Z_j^0$, for every neighborhood $U \ni x$ and for every holomorphic embedding $\iota : U \to \mathbb{D}^k \times \mathbb{D}^N$, $N \in \mathbb{N}$, such that $\iota(x) = 0$ and the projection

$$pr_{\mathbb{D}^k} : \iota\left(U \cap \bigcup_j Z_j^0\right) \to \mathbb{D}^k$$

is proper, the map

$$e_k : \mathscr{C}_k(X) \times \mathbb{D}^k \to \mathscr{C}_0(\mathbb{D}^N)$$

defined by

$$e_k(Z, t) = \sum m_j pr_{\mathbb{D}^N}\left(\iota(Z_j \cap U) \cap (\{t\} \times \mathbb{D}^N)\right), \text{ where } Z = \sum m_j Z_j,$$

is holomorphic on an open and dense subset of a neighborhood of Z^0, where the topology of $\mathscr{C}_k(X)$ is defined in an obvious manner.

It was proved in [10] that $\mathscr{C}_{q-1}(X)$ is Stein if (X, φ) is q-complete.[3] A 1-convex exhaustion function Φ on a connected component of $\mathscr{C}_{q-1}(X)$ is defined by $\Phi(x) = \max_y \varphi(y)$, where y runs through the points of $\bigcup_j Z_j$ for which there exist m_j such that $x = \sum_j m_j Z_j$. The union $\bigcup_j Z_j$ will be called the *support of* $\sum m_j Z_j$. The same method seems to work to show that $\mathscr{C}_{q-k}(X)$ is k-complete if X is q-complete.

The construction of Φ is first due to Norguet and Siu [129], who proved the following as a natural generalization of [4].

Theorem 4.2 *Let X be a q-convex domain in a projective algebraic manifold. Then $\mathscr{C}_{q-1}(X)$ is holomorphically convex. If moreover $H^q(X, \mathscr{F}) = 0$ for all coherent analytic sheaf $\mathscr{F}$ on X, then $\mathscr{C}_{q-1}(X)$ is Stein.*

---

[1] See [30] for some explicit expressions of Chow varieties.

[2] *Holomorphically convex* for reduced complex spaces means the same as in Theorem 1.4.

[3] See also [11–13].

Note that there exists a 1-convex space X of dimension 3 such that $\mathscr{C}_1(X)$ is not holomorphically convex (cf. [10]). We note that the topological countablity of $\mathscr{C}_k(X)$ follows from a basic result of Fujiki [66] on that of the Douady spaces. Fujiki [67] also proved that $\mathscr{C}_{n-1}(X)$ has projective algebraic components for every compact complex space X.

In [132], the original method of [3] was refined by the method of L^2-estimates for the $\bar{\partial}$-operator to show the following.

Theorem 4.3 *Let M be a q-complete manifold with a q-convex exhaustion function φ and let $(\gamma_\mu)_{\mu \in \mathbb{N}}$ be a sequence in $\mathscr{C}_{q-1}(M)$. Suppose that one can choose points x_μ from the supports of γ_μ in such a way that x_μ does not accumulate to any point of M. Then there exists a complete Hermitian metric g on M and a C^∞-smooth convex increasing function $\tau : \mathbb{R} \to \mathbb{R}$ and $K > 0$ such that, for any sequence $c = \{c_\mu\} \subset \mathbb{C}$ satisfying*

$$\|c\|_\tau := \sum_{\mu=1}^{\infty} |c_\mu \exp(-\tau(\varphi(x_\mu)))| < \infty$$

one can find a C^∞-smooth $\bar{\partial}$-closed $(q-1, q-1)$-form ω on M fulfilling

$$\|\omega\|_\tau := \int_M |\omega|_g \exp(-\tau \circ \varphi) dv_g \le K \|c\|_\tau$$

and

$$\int_{\gamma_\mu} \omega = c_\mu \text{ for every } \mu \in \mathbb{N}.$$

Here $|\cdot|_g$ and dv_g denote respectively the length and the volume form with respect to g and $\int_{\sum_j m_j z_j} := \sum_j m_j \int_{z_j}$.

We note that the L^2-method was first applied in several complex variables to the sheaf cohomology theory by Kodaira [106, 107] to prove a projective embedding theorem for compact complex manifolds admitting positive line bundles. Andreotti and Vesentini [6, 7] first noted that Kodaira's method is applicable to prove basic existence theorems by Oka and Cartan on 1-complete manifolds as well as a generalization of Kodaira's embedding theorem to noncompact manifolds. Similar results on the $\bar{\partial}$-equation involving deeper analysis have been developed by Kohn [109, 110] and Hörmander [94, 95] (see also [111]), realizing the ideas of Bergman [23] and Garabedian and Spencer [71] by extending the work of Morrey [121]. In view of these classical works, it should not be too hard to formulate and prove the q-convex version of Theorem 4.3.

4.3   *q*-Convex Domains in Projective Spaces

Oka's solution of the Levi problem was generalized by R. Fujita [70] and A. Takeuchi [164] for Riemann domains over $\mathbb{CP}^n$. They proved that locally pseudoconvex Riemann domains over $\mathbb{CP}^n$ are either $\mathbb{CP}^n$ or Stein. For the proof, Fujita took advantage of the fact that $\mathbb{CP}^n$ is covered by $n + 1$ copies of $\mathbb{C}^n$ and Takeuchi generalized Oka's lemma in [135] to estimate the lower bound of the eigenvalues of $-\partial\bar{\partial}\log\delta$ for the distance δ to the boundary of locally Stein domains in $\mathbb{CP}^n$ with respect to the Fubini-Study metric. These methods were applied, respectively, by Ueda [165] to solve the Levi problem on Grassmann varieties and by Schwarz [152] and Matsumoto [117, 118] to prove that locally q-complete domains with C^2-smooth boundary in $\mathbb{CP}^n$ are q-convex. Here, a domain D in a complex manifold is said to be *locally q-complete* if for every point $x \in \partial D$ there is a neighbourhood U of x such that $U \cap D$ is q-complete.

An interesting example of q-convexity is given by complex submanifolds Y of $\mathbb{CP}^n$. Barth [15, 16] initiated the studies on $\mathbb{CP}^n \setminus Y$ motivated by questions in algebraic geometry raised by Hartshorne [85]. Recall that a complex analytic set $A \subset \mathbb{CP}^n$ is called a *complete intersection* if the ideal of $\mathbb{C}[z_0, z_1, \ldots, z_n]$ defining A is generated by a set whose cardinality is codim$A$. Not all submanifolds are complete intersections but there are results suggesting that a submanifold Y is a complete intersection if codim$Y < \frac{n}{3}$ (cf. [86]). See also [17, 57, 58].

Definition 4.2 An analytic set $A \subset \mathbb{CP}^n$ is called a *set-theoretical complete intersection* (SCI, for short) if A is the intersection of hypersurfaces $S_1, \ldots, S_q$ with $q = $ codim$A$.

It is easy to see that the complement of a q-codimensional SCI is q-complete. Not every q-codimensional submanifold Y is an SCI. For instance, SCIs of codimension q in $\mathbb{CP}^n$ must have trivial first Betti number if $n \geq q + 2$. Indeed, if codim$A = q$, one has $H^k(\mathbb{CP}^n \setminus A, \mathbb{C}) = 0$ for all $k \geq n + q$ since $\mathbb{CP}^n \setminus A$ is then covered by q Stein open subsets, so that $b_1(A) = 0$ follows from the exact sequence

$$H^1(\mathbb{CP}^n, \mathbb{C}) \to H^1(A, \mathbb{C}) \to H_c^2(\mathbb{CP}^n \setminus A, \mathbb{C}),$$

where H^r (resp. H_c^r) denote the r-th de Rham cohomology groups (resp. with compact support), combined with the Poincaré duality $H^r(\mathbb{CP}^n \setminus Y, \mathbb{C}) \cong H_c^{2n-r}(\mathbb{CP}^n \setminus Y, \mathbb{C})$ and $H^1(\mathbb{CP}^n, \mathbb{C}) = 0$ for any n. There exist 2-dimensional complex tori embedded in $\mathbb{CP}^4$ (cf. [93]). Hence by the topological restriction $b_1(A) = 0$ for A to be an SCI, 2-tori in $\mathbb{CP}^4$ are not SCIs. A long-standing open question is whether or not every connected complex curve in $\mathbb{CP}^n$ is an SCI. The corresponding question for the $(n - 1)$-completeness of the complement is also highly delicate. It is known from Colţoiu and Diederich [44] that $\mathbb{CP}^n \setminus C$ is locally $(n - 1)$-complete for any complex curve C, but Colţoiu [40] found a locally 2-complete domain in $\mathbb{CP}^3$ which is not 2-convex.

Anyway, Barth first proved the following, which is a special case of the above-mentioned result of Schwarz and Matsumoto.

Theorem 4.4 $\mathbb{CP}^n \setminus Y$ *is q-convex if the codimension of Y is q.*

In the above situation, $\mathbb{CP}^n \setminus Y$ would be q-complete if the Fubini–Study distance to Y is C^2-smooth outside a finite set. However, it is not the case in general, as was explained above (see also [28]).

A bound for the k-completeness of $\mathbb{CP}^n \setminus Y$ was obtained by Diederich and Fornæss [50, 51].

Theorem 4.5 $\mathbb{CP}^n \setminus Y$ *is $\tilde{q}$-complete with* $\tilde{q} = n - \left[\frac{n}{q}\right] + 1$ *if the codimension of Y is q. Here,* $[\alpha] := \max\{m : m \leq \alpha \text{ and } m \in \mathbb{Z}\}$.

The proof of Theorem 4.5 is based on the fact that any *q-convex function with corners* (a continuous function which is locally the maximum of finitely many q-convex functions) can be approximated, for any given width of positive continuous function, by C^∞-smooth $\tilde{q}$-convex functions, where $\tilde{q} = n - \left[\frac{n}{q}\right] + 1$. A *q-convex space with corners* is defined to be a complex space equipped with an exhaustion function which is q-convex with corners. The notion of q-convexity with corners was introduced by Grauert [76]. M. Peternell [145] showed a result which implies that $\mathbb{CP}^n \setminus Y$ is $(2q-1)$-complete, which is stronger than Theorem 4.5 if $1 < q < \frac{n}{2}$. A better bound is known for the cohomology vanishing with coefficients in the algebraic sheaves (cf. [36]).

Since the holomorphic bisectional curvature of $\mathbb{CP}^n$ is positive, the normal bundle N_Y of Y is positive. Schneider [151] showed that this positivity property of N_Y suffices to show Theorem 4.4 (see also [64, 65]). In this case we get:

Theorem 4.6 *If $\mathcal{F}$ is a coherent analytic sheaf on $X = \mathbb{CP}^n \setminus Y$, then the cohomology groups $H^j(X, \mathcal{F})$ have finite dimension for $j \geq q$.*

For this finite-dimensionality assertion, the smoothness of Y is essential. In fact, remove from $\mathbb{C}^4$ the union of two planes intersecting transversally. Then the rest A of the domain in $\mathbb{CP}^4 \supset \mathbb{C}^4$ obtained in this way is of codimension 2 and $\dim H^2(\mathbb{CP}^4 \setminus A, \mathcal{O}) = \infty$, where $\mathcal{O}$ denotes the structure sheaf (cf. [77]).

On the other hand, Hartshorne [85] proved the following.

Theorem 4.7 *Let C be any connected algebraic curve in $\mathbb{CP}^n$ and let $\mathcal{F} \to \mathbb{CP}^n$ be any coherent algebraic sheaf. Then the $(q-1)$-th algebraic cohomology group of $\mathbb{CP}^n \setminus C$ with coefficients in $\mathcal{F}$ vanishes.*

If C is an arbitrary algebraic curve in $\mathbb{CP}^n$, then $\dim H^{n-1}(\mathbb{CP}^n \setminus C, \mathcal{F}) < \infty$, as was shown by M. Peternell [144]. Further, for such C with k connected components, Colţoiu [41] showed that

$$\dim H^{n-1}(\mathbb{CP}^n \setminus C, \mathcal{G}) = (k-1) \dim H^0(\mathbb{CP}^n \setminus C, \mathrm{Hom}(\mathcal{G}, \mathcal{K})) < \infty,$$

holds for any coherent analytic sheaf $\mathscr{G}$ over $\mathbb{CP}^n \setminus C$, where $\mathscr{K}$ denotes the canonical sheaf of $\mathbb{CP}^n$. In particular, $\dim H^{n-1}(\mathbb{CP}^n \setminus C, \mathscr{G}) = 0$ for any coherent analytic sheaf $\mathscr{G}$ if C is connected, as was suggested by Theorem 4.7. We point out that Hartshorne considers algebraic cohomology groups, while Colţoiu and Peternell deal with analytic cohomology groups.

At this point a question posed by Andreotti and Grauert [2] must be mentioned:

Given a complex analytic space X, is the vanishing of $H^k(X, \mathscr{F})$ for all $k \geq q$ and for all coherent analytic sheaves $\mathscr{F}$ over X equivalent to the q-completeness of X?

Replacing the vanishing by the finite-dimensionality, one has the corresponding conjecture for the characterization of q-convex spaces. Affirmative partial answers have been found in [48, 49, 118, 119] and [169].

Remark 4.1 Andreotti–Grauert's conjecture seems to be based on an observation that the union of q Stein open sets in a complex manifold is *cohomologically q-complete* in the sense that the k-th cohomology groups with coefficients in coherent analytic sheaves are zero for $k \geq q$. Actually, it was proved in [145] that a complex manifold is q-complete if it is covered by 1-complete open sets (U_i, φ_i), $i = 1, \ldots, q$. Since the proof is quite elementary, it may be worthwhile to put it here. For simplicity, we restrict ourselves to the case of manifolds, but the generalization to analytic spaces is immediate.

Proof of the q-Completeness of the Union of U_i If $q = 1$, there is nothing to prove. Suppose that the assertion is true for $q < r$ and let M be a complex manifold covered by 1-complete open sets $U_1, \ldots, U_r$. By the induction hypothesis $U := \bigcup_{i=1}^{r-1} U_i$ is $(r-1)$-complete. Let φ be an $(r-1)$-convex function on U and let ψ be a strictly plurisubharmonic function on U_r. Then it is easy to verify that $-\log(e^{-\varphi} + e^{-\psi})$ is r-convex on $U \cap U_r$. In fact

$$\partial\bar{\partial}(-\log(e^{-\varphi} + e^{-\psi})) = \frac{-\partial\bar{\partial}(e^{-\varphi} + e^{-\psi})}{e^{-\varphi} + e^{-\psi}} + \frac{\partial(e^{-\varphi} + e^{-\psi})\bar{\partial}(e^{-\varphi} + e^{-\psi})}{(e^{-\varphi} + e^{-\psi})^2}$$

$$= \frac{(e^{-\varphi} + e^{-\psi})(e^{-\varphi}\partial\bar{\partial}\varphi + e^{-\psi}\partial\bar{\partial}\psi) - e^{-\varphi}e^{-\psi}(\partial\varphi - \partial\psi)(\bar{\partial}\varphi - \bar{\partial}\psi)}{(e^{-\varphi} + e^{-\psi})^2}.$$

Note that, by composing some convex increasing function to φ and ψ in advance, one may assume that $e^{-\varphi}$ and $e^{-\psi}$ are smoothly extendable on M in such a way that they are zero on $M \setminus U$ and $M \setminus U_r$, respectively. Hence, for any C^∞-smooth strictly plurisubharmonic exhaustion functions φ_i on U_i, one can define inductively an r-convex exhaustion function on M. $\square$

In view of Theorems 4.2, 4.4, and 4.7 also suggests that $\mathscr{C}_{n-2}(\mathbb{CP}^n \setminus C)$ is Stein for any connected algebraic curve $C \subset \mathbb{CP}^n$. Compared to the questions of complete intersection and q-completeness, the Steinness of cycle spaces has a simpler aspect as follows.

Proposition 4.1 *Let $Y \subset \mathbb{CP}^n$ be a complex analytic set of codimension q and let X be a q-convex domain in $\mathbb{CP}^n \setminus Y$. Then $\mathscr{C}_{q-1}(X)$ is Stein.*

From this, we derive:

Corollary 4.1 *Let $C \subset \mathbb{CP}^n$ be a (not necessarily connected) smooth algebraic curve. Then $\mathscr{C}_{n-2}(\mathbb{CP}^n \setminus C)$ is Stein.*

Proof of Proposition 4.1 If $\mathscr{C}_{q-1}(X)$ contained a positive-dimensional compact analytic set, X would contain a compact analytic set of dimension $\geq q$, which contradicts a well-known fact that

$$\dim(Y \cap Z) \geq \dim Y + \dim Z - n$$

holds for any subvarieties Y and Z in $\mathbb{CP}^n$ (cf. [87, p.48, Theorem 7.2]).   □

For any domain $D \subset \mathbb{CP}^n$ let us put

$$\mathscr{C}_k^d(D) = \left\{ \sum m_j Z_j \in \mathscr{C}_k(D) : \sum m_j \deg Z_j = d \right\}.$$

Since the Picard varieties of Grassmann varieties are known to be $\mathbb{Z}$, which follows from the homology of the Grassmannian in terms of Schubert cycle and from the Hodge decomposition (cf. [81]), one has the following.

Proposition 4.2 *The set $\mathscr{C}_{q-1}^1(\mathbb{CP}^n \setminus A)$ is Stein for any q-codimensional analytic set $A \subset \mathbb{CP}^n$.*

Proposition 4.2 may be regarded as a reason to suspect that $\mathscr{C}_{q-1}(\mathbb{CP}^n \setminus A)$ is Stein for *all* q-codimensional A. Recall that this is true for $q = 1$ and $q = n$. The case $q = 1$ is by Chow's theorem on the algebraicity of analytic sets in $\mathbb{CP}^n$ and the latter follows from a theorem of Greene and Wu in [80] asserting the n-completeness of noncompact and connected complex manifolds (see also [131] and [47]). To be more precise, if $q = 1$ or $q = n$, $X = \mathbb{CP}^n \setminus A$ is q-complete if and only if $\mathscr{C}_{q-1}(X)$ is Stein. For some classes of q-complete domains, hyperbolicity questions naturally arises (cf. [96]). Concerning the properties of $\mathscr{C}_p(X)$ for $p \neq q - 1$, it was proved by Barlet and Vâjâitu [14] that $\mathscr{C}_q(X)$ is 2-complete with corners if X is a $(q + r)$-complete Kähler space for some non-negative integer r such that $H^k(X, \mathscr{F}) = 0$ $(k \geq q)$ holds for any coherent analytic sheaf $\mathscr{F} \to X$. It seems also reasonable to ask whether or not the connected components of $\mathscr{C}_{q-k}(X)$ are k-convex if X is a q-convex Kähler space with $q \geq k$.

4.4 *q*-Convexity in Analytic Families

We start with a definition.

Definition 4.3 Let X be a reduced complex space with a holomorphic map π onto a complex space T whose fibers are connected. We say X is *q-convex* (resp. *locally q-complete*) *over* T if for any $t \in T$ there exist a neighborhood $U \ni t$ and a C^2-smooth function $\varphi : \pi^{-1}(U) \to [0, \infty)$ satisfying the following properties:

(1) $\pi|_{\varphi^{-1}([0,c])}$ is proper for all $c \in \mathbb{R}$.
(2) φ is q-convex on $\varphi^{-1}((1, \infty))$ (resp. on $\pi^{-1}(U)$).[4]

π is said to be a *q-convex* map (resp. a *locally q-complete* map) if X is q-convex (resp. locally q-complete) over T. π is called a *locally holomorphically convex* map (resp. a *locally Stein* map) if every point of T has a neighborhood whose preimage by π is holomorphically convex (resp. Stein).

By a theorem of Knorr and Schneider [105], we know that 1-convex maps are locally holomorphically convex. By the construction of the exhaustion function Φ in Theorem 4.2, the following relative version of Theorem 4.2 follows immediately from this theorem.

Theorem 4.8 *Let X be a locally closed analytic set in $\mathbb{CP}^n$ and let T be a Stein space. Then, for every q-convex map $\pi : X \to T$, the induced map $\mathscr{C}_{q-1}(X) \to T$ is a locally holomorphically convex map.*

We say that $\pi : X \to T$ is an *analytic family* of complex manifolds if X and T are connected complex manifolds and π is a submersion onto T whose fibers are connected and diffeomorphic to each other. Deformation theory of complex structures describes how the complex structures of the fibers $X_t := \pi^{-1}(t)$ change under various circumstances. Kodaira and Spencer [108] defined the derivative of the complex structures of X_t at $t \in T$ as a map from the tangent space of T at t with values in $H^{0,1}(X_t, \Theta_{X_t})$ (the Kodaira–Spencer map), where Θ_{X_t} denotes the holomorphic tangent bundle of X_t. The most basic result in this context is the rigidity theorem which says that π is locally trivial, i.e. $\pi^{-1}(U) \cong X_t \times U$ for each t for a sufficiently small neighborhood $U \ni t$, if π is proper and the Kodaira–Spencer map is zero on T.

Based on this rigidity theorem, Fischer and Grauert [59] showed that π is locally trivial if X_t are biholomorphically equivalent to each other. If an analytic family $\pi : X \to T$ is locally Stein, no simple general criterion is known for the local triviality. Nevertheless there are remarkable positive results for the families of Riemann surfaces which deserve to be better known. Yamaguchi [168] proved that a Stein analytic family is locally trivial if the fibers are isomorphic to each

---

[4] Since the above $\pi^{-1}(U)$ need not be q-convex, we shall not use the terminology "locally q-convex".

other, provided that they are equivalent neither to $\mathbb{D}$ nor $\mathbb{D} \setminus \{0\}$.[5] The result is a generalization of Nishino's rigidity theorem for Stein families of $\mathbb{C}$ (cf. [126]). It is expected that Nishino's theorem is true for analytic families of $\mathbb{C}^n$ for arbitrary n, but the proof seems to be far-reaching. Nevertheless the following can be proved at least (cf. [133, Theorem 4.3]).

Theorem 4.9 *Let $n \geq 2$ and let π be a holomorphic submersion from a complex manifold M onto a Stein manifold N such that $H^2(N, \mathbb{Z}) = 0$. Assume that there exists a complete Kähler metric on M, $\pi^{-1}(t) \cong \mathbb{C}^n$ for all $t \in N$ and that there exists a proper holomorphic embedding*

$$\sigma : N \times \{z = (z_1, \ldots, z_n) \in \mathbb{C}^n : \min\{|z_1|, \ldots, |z_n|\} \leq 1\} \hookrightarrow M$$

satisfying $\pi \circ \sigma(t, z) \equiv t$. Then there exists a biholomorphic map F from M to $N \times \mathbb{C}^n$ commuting with π and the projection to N such that the image of σ is mapped by F onto $N \times \{z = (z_1, \ldots, z_n) \in \mathbb{C}^n : \min\{|z_1|, \ldots, |z_n|\} \leq 1\}$.

The L^2-method is available for the proof. The role of the L^2-condition is illustrated as follows.

Based on the fact that $f(z)dz$ is a constant multiple of $z^{-2}dz$ if and only if $f \in O(\mathbb{C} \setminus \{z = 0\})$ and

$$\int_{\mathbb{C} \setminus \{z = 0\}} e^{-3 \log^+ (1/|z|)} |f(z)|^2 < \infty$$

and that any holomorphic 1-form u on $\mathbb{C} \setminus \{0\}$ satisfying

$$i \int_{\mathbb{C} \setminus \{0\}} e^{-5 \log^+ (1/|z|)} u \wedge \bar{u} < \infty$$

is of the form

$$\left(\frac{a}{z^2} + \frac{b}{z^3} \right) dz, \text{ where } a, b \in \mathbb{C},$$

one has a univalent map from $\mathbb{C} \setminus \{0\}$ to $\mathbb{C}$ by taking the ratio v/u of an L^2-holomorphic 1-form u with respect to the weight $3 \log^+(1/|z|)$ and an L^2-holomorphic 1-form v with respect to the weight $5 \log^+ (1/|z|)$.

A direct consequence of Theorem 4.9 is the following (cf. [133, Theorem 0.2]).

Theorem 4.10 *An analytic family $\pi : X \to T$ is locally trivial if $X_t \cong \mathbb{CP}^n \setminus \{x\}$, for all $t \in T$ and π is an n-convex map, where $x \in \mathbb{CP}^n$.*

---

[5] The proof is given only for the case where the fibers are of finite topological type "for simplicity", but generalization to the case of infinite topological type is immediate.

The proof is done by showing that the map $\mathscr{C}^1_{n-1}(X) \to T$ associated with π, with the obvious definition of $\mathscr{C}^1_{n-1}(X) \to T$, is a locally trivial family of $\mathbb{C}^n$ with canonical biholomorphisms $\mathscr{C}^1_{n-1}(X_t) \to \mathbb{C}^n$.

Corollary 4.2 *For any $\mathbb{C}^n$-valued continuous function f on a complex manifold T, f is holomorphic if and only if the projection $T \times \mathbb{C}^n \setminus \{(t, f(t)); t \in T\} \to T$ is n-convex.*

Proof of Corollary 4.2 By Theorem 4.9, there exists a holomorphic embedding ι of $T \times \mathbb{C}^n \setminus \{(t, f(t)); t \in T\}$ into $T \times (\mathbb{CP}^n \setminus \{x\})$ for some $x \in \mathbb{CP}^n$ commuting with the projections. By extending ι fiberwise, one has a holomorphic embedding of $T \times \mathbb{C}^n$ into $T \times \mathbb{CP}^n$. Hence, the preimage of $T \times \{x\}$ is analytic because so is the graph of f. □

Corollary 4.2 is also included in Theorem 3.7 from Chap. 3 (or [139, 140]) by reducing the general case to the case $n = 1$ noticing that the graph of each component of f has a 1-convex complement. When $n = 1$ the assertion is the Hartogs' theorem [84].

It is likely that Theorem 4.10 is related to the question whether or not the analytic family arising as the universal covering of that of Hopf surfaces is locally trivial. In connection to this problem, it might be worthwhile to mention the works of Colţoiu and Vâjâitu [45] and Miyazawa [120] on the local n-completeness of covering spaces of analytic families of compact complex manifolds (see also [43]).

References

1. Alexandroff, A.D.: Almost everywhere existence of the second differential of a convex function and some properties of convex surfaces connected with it. Leningrad State Univ. Annals [Uchenye Zapiski] Math. Ser. **6**, 3–35 (1939) (Russian)
2. Andreotti, A., Grauert, H.: Théorème de finitude pour la cohomologie des espaces complexes. Bull. Soc. Math. France **90**, 193–259 (1962)
3. Andreotti, A., Norguet, F.: Problème de Levi et convexité holomorphe pour les classes de cohomologie. Ann. Scuola Norm. Sup. Pisa Cl. Sci. (3) **20**, 197–241 (1966)
4. Andreotti, A., Norguet, F.: La convexité holomorphe dans l'espace analytique des cycles d'une variété algébrique. Ann. Scuola Norm. Sup. Pisa Cl. Sci. (3) **21**, 31–82 (1967)
5. Andreotti, A., Norguet, F.: Cycles of algebraic manifolds and $\partial\bar{\partial}$-cohomology. Ann. Scuola Norm. Sup. Pisa Cl. Sci. (3) **25**, 59–114 (1971)
6. Andreotti, A., Vesentini, E.: Sopra un teorema di Kodaira. Ann. Scuola Norm. Sup. Pisa Cl. Sci. (3) **15**, 283–309 (1961)
7. Andreotti, A., Vesentini, E.: Les théorèmes fondamentaux de la théorie des espaces holomorphiquement complets. Cahiers Sémin. Topol. et Géom. Différent. C. Ehresmann, Fac. Sci. Paris, Vol. IV, 1–31 (1962–1963)
8. Baouendi, M.S., Ebenfelt, P., Rothschild, L. P.: Real submanifolds in complex space and their mappings. Princeton Math. Series, vol. 47. Princeton University Press (1999)
9. Barlet, D.: Espace analytique réduit des cycles analytiques complexes compacts d'un espace analytique complexe de dimension finie. Fonctions de plusieurs variables complexes, II (Sém. François Norguet, 1974–1975), pp. 1–158. Lecture Notes in Math., Vol. 482, Springer, Berlin, Heidelberg (1975)
10. Barlet, D.: Convexité de l'espace des cycles. Bull. Soc. Math. France **106**, no. 4, 373–397 (1978)
11. Barlet, D., Magnússon, J.: Cycles analytiques complexes. I. Théorèmes de préparation des cycles. Cours Spécialisés, **22**. Société Mathématique de France, Paris (2014)
12. Barlet, D., Magnússon, J.: Complex analytic cycles I: Basic results on complex geometry and foundations for the study of cycles. Grundlehren der mathematischen Wissenschaften **356** (2020)
13. Barlet, D., Magnússon, J.: Complex analytic cycles II. in press and to appear in 2022
14. Barlet, D., Vâjâitu, V.: Convexity properties for cycle spaces. Michigan Math. J. **50**(1), 57–70 (2002)
15. Barth, W.: Der Abstand von einer algebraischen Mannigfaltigkeit im komplex-projektiven Raum. Math. Ann. **187**, 150–162 (1970)

16. Barth, W.: Transplanting cohomology classes in complex-projective space. Am. J. Math. **92**, 951–967 (1970)

17. Barth, W., Van de Ven, A.: A decomposability criterion for algebraic 2-bundles on projective spaces. Invent. Math. **25**, 91–106 (1974)

18. Basener, R.F.: Nonlinear Cauchy-Riemann equations and q-pseudoconvexity. Duke Math. J. **43**(1), 203–213 (1976)

19. Bedford, E., Kalka, M: Foliations and complex Monge-Ampère equations. Commun. Pure Appl. Math. **30**(5), 543–571 (1977)

20. Behnke, H.: Der Kontinuitätssatz und die Regulärkonvexität. Math. Ann. **11**, 392–397 (1936)

21. Behnke, H., Stein, K.: Konvergente Folgen nichtschlichter Regularitätsbereiche. Ann. Mat. Pura Appl. **28**, 317–326 (1949)

22. Behnke, H., Sommer, F.: Analytische Funktionen mehrerer komplexer Veränderlichen. Über die Voraussetzungen des Kontinuitätssatzes. Math. Ann. **121**, 356–378 (1950)

23. Bergman, S.: Über die Kernfunktion eines Bereiches und ihr Verhalten am Rande. I. J. Reine Angew. Math. **169**, 1–42 (1933), and **172**, 89–128 (1934)

24. Bremermann, H.J.: Über die Äquivalenz der pseudokonvexen Gebiete und der Holomorphiegebiete im Raum von n komplexen Veränderlichen. Math. Ann. **128**, 63–91 (1954)

25. Bremermann, H.J.: On the conjecture of the equivalence of the plurisubharmonic functions and the Hartogs functions. Math. Ann. **131**, 76–86 (1956)

26. Bremermann, H.J.: Die Charakterisierung Rungescher Gebiete durch plurisubharmonische Funktionen. Math. Ann. **136**, 173–186 (1958)

27. Bremermann, H.J.: On a generalized Dirichlet problem for plurisubharmonic functions and pseudo-convex domains. Characterization of Šilov boundaries. Trans. Am. Math. Soc. **91**, 246–276 (1959)

28. Buchner, M., Fritzsche, K., Sakai, T.: Geometry and cohomology of certain domains in the complex projective space. J. Reine Angew. Math. **323**, 1–52 (1981)

29. Bungart, L.: Piecewise smooth approximations to q-plurisubharmonic functions. Pac. J. Math. **142**(2), 227–244 (1990)

30. Bürgisser, P., Kohn, K., Lairez, P., Sturmfels, B.: Computing the Chow variety of quadratic space curves. Mathematical aspects of computer and information sciences, pp. 130–136. Lecture Notes in Comput. Sci., vol. 9582. Springer, Cham (2016)

31. Busemann, H., Feller, W.: Krümmungseigenschaften Konvexer Flächen. Acta Math. **66**(1), 1–47 (1936)

32. Cartan, H.: Fonctions analytiques de plusieurs variables complexes. Paris: Séminaire E.N.S. (1951–1952)

33. Cartan, H., Serre, J.P.: Un théorème de finitude concernant les variétés analytiques compactes. C. R. Acad. Sci., Paris **237**, 128–130 (1953)

34. Cartan, H., Serre, J.-P.: Une théorème de finitude concernant les variétés analytiques compactes. C.R. Acad. Sci. Paris **237**, 128–130 (1953)

35. Cartan, H., Thullen, P.: Zur Theorie des Singularitäten der Funktionen mehrerer komplexen Veränderlichen. Math. Ann. **106**, 617–647 (1932)

36. Chiriacescu, G., Colţoiu, M., Joiţa, C.: Analytic cohomology groups in top degrees of Zariski open sets in $\mathbb{P}^n$. Math. Z. **264**(3), 671–677 (2010)

37. Chirka, E.M.: Complex Analytic Sets. volume 46 of Mathematics and Its Applications (Soviet Series). Kluwer Academic Publishers Group, Dordrecht (1989)

38. Chirka, E.M.: Levi and Trépreau theorems for continuous graphs. Tr. Mat. Inst. Steklova **235**(Anal. i Geom. Vopr. Kompleks. Analiza), 272–287 (2001)

39. Colţoiu, M.: Traces of Runge domains on analytic subsets. Math. Ann. **290**(3), 545–548 (1991)

40. Colţoiu, M.: A counter-example to the q-Levi problem in $\mathbb{P}^n$. J. Math. Kyoto Univ. **36**(2), 385–387 (1996)

41. Colţoiu, M.: On Barth's conjecture concerning $H^{n-1}(\mathbb{P}^n \setminus A, F)$. Nagoya Math. J. **145**, 99–123 (1997)

42. Colţoiu, M.: q-convexity. A survey. Complex analysis and geometry (Trento, 1995), pp. 83–93, Pitman Res. Notes Math. Ser., vol. 366. Longman, Harlow (1997)
43. Colţoiu, M., Diederich, K.: On the coverings of proper families of 1-dimensional complex spaces. Michigan Math. J. **47**(2), 369–375 (2000)
44. Colţoiu, M., Diederich, K.: Convexity properties of analytic complements in Stein spaces. Proceedings of the Conference in Honor of Jean-Pierre Kahane (Orsay, 1993). J. Fourier Anal. Appl. Special Issue, 153–160 (1995)
45. Colţoiu, M., Vâjâitu, V.: On n-completeness of covering spaces with parameters. Math. Z. **237**, 815–831 (2001)
46. Crandall, M.G., Ishii, H., Lions, P.L.: User's guide to viscosity solutions of second order partial differential equations. Bull. Am. Math. Soc. (N. S.) **27**, 1–67 (1992)
47. Demailly, J.-P.: Cohomology of q-convex spaces in top degrees. Math. Z. **204**(2), 283–295 (1990)
48. Demailly, J.-P.: A converse to the Andreotti-Grauert theorem. Ann. Fac. Sci. Toulouse Math. (6) **20**, 123–135 (2011)
49. Demailly, J.-P., Peternell, T., Schneider, M.: Holomorphic line bundles with partially vanishing cohomology. In Proceedings of the Hirzebruch 65 conference on algebraic geometry, Israel Mathematical Conference Proceedings, vol. 9 (Bar-Ilan University, Ramat Gan), pp. 165–198 (1996)
50. Diederich, K., Fornaess, J.E.: Smoothing q-convex functions and vanishing theorems. Invent. Math. **82**(2), 291–305 (1985)
51. Diederich, K., Fornaess, J.E.: Smoothing q-convex functions in the singular case. Math. Ann. **273**(4), 665–671 (1986)
52. Diederich, K., Sukhov, A.: Plurisubharmonic exhaustion functions and almost complex Stein structures. Michigan Math. J. **56**(2), 331–355 (2008)
53. Dieu, Nguyen Quang: q-plurisubharmonicity and q-pseudoconvexity in $\mathbb{C}^n$. Publ. Mat. **50**, 349–369 (2006)
54. Docquier, F., Grauert, H.: Levisches Problem und Rungescher Satz für Teilgebiete Steinscher Mannigfaltigkeiten. Math. Ann. **140**, 94–123 (1960)
55. Eastwood, M.G., Suria, G.V.: Cohomologically complete and pseudoconvex domains. Commun. Math. Helv. **55**(3), 413–426 (1980)
56. El Mir, H.: Sur le prolongement des courants positifs fermes. Acta Math. **153**, 1–45 (1984)
57. Erman, D., Sam, S.V., Snowden, A.: Strength and Hartshorne's conjecture in high degree. Math. Z. **297**, 1467–1471 (2021)
58. Faltings, G.: Ein Kriterium für vollständige Durchschnitte. Invent. Math. **62**, 393–401 (1981)
59. Fischer, W., Grauert, H.: Lokal-triviale Familien kompakter komplexer Mannigfaltigkeiten. Nachr. Akad. Wiss. Göttingen Math.-Phys. Kl. II 1965, 89–94 (1965)
60. Fornæss, J.E., Narasimhan, R.: The Levi problem on complex spaces with singularities. Math. Ann. **248**, 47–72 (1980)
61. Fornæss, J.E., Stensønes, B.: Lectures on counterexamples in several complex variables. AMS Chelsea Publishing, Providence (2007)
62. Forstnerič, F.: Stein manifolds and holomorphic mappings. The homotopy principle in complex analysis. Ergebnisse der Mathematik und ihrer Grenzgebiete. 3. Folge. A Series of Modern Surveys in Mathematics, vol. 56. Springer, Cham (2017)
63. Freeman, M.: Local complex foliation of real submanifolds. Math. Ann. **209**, 1–30 (1974)
64. Fritzsche, K.: q-konvexe Restmengen in kompakten komplexen Mannigfaltigkeiten. Math. Ann. **221**(3), 251–273 (1976)
65. Fritzsche, K.: Pseudoconvexity properties of complements of analytic subvarieties. Math. Ann. **230**(2), 107–122 (1977)
66. Fujiki, A.: Countability of the Douady space of a complex space. Japan. J. Math. (N.S.) **5**(2), 431–447 (1979)
67. Fujiki, A.: Projectivity of the space of divisors on a normal compact complex space. Publ. Res. Inst. Math. Sci. **18**(3), 1163–1173 (1982)
68. Fujita, O.: Sur les familles d'ensembles analytiques. J. Math. Soc. Japan **16**, 379–405 (1964)

69. Fujita, O.: On the equivalence of the q-plurisubharmonic functions and the pseudoconvex functions of general order. Ann. Reports of Graduate School of Human Culture, Nara Women's Univ. **7**, 77–81 (1992)

70. Fujita, R.: Domaines sans point critique intérieur sur l'espace projectif complexe. J. Math. Soc. Jpn. **15**, 443–473 (1963)

71. Garabedian, P.R., Spencer, D.C.: Complex boundary value problems. Trans. Am. Math. Soc. **73**, 223–242 (1952)

72. Grauert, H.: On Levi's problem and the imbedding of real-analytic manifolds. Ann. Math. (2) **68**, 460–472 (1958)

73. Grauert, H.: Une notion de dimension cohomologique dans la théorie des espaces complexes. Bull. Soc. Math. France **87**, 341–350 (1959)

74. Grauert, H.: Über Modifikationen und exzeptionelle analytische Mengen. Math. Ann. **146**, 331–368 (1962)

75. Grauert, H.: Die Bedeutung des Levischen Problems für die analytische und algebraische Geometrie. Proceedings of the International Congress of Mathematicians, Bd. 1. ICM, Stockholm, pp. 86–101 (1963)

76. Grauert, H.: Kantenkohomologie. Compositio Mathematica, **44**(1–3), 79–101 (1981)

77. Grauert, H.: Theory of q-convexity and q-concavity. Complex analysis. Several variables, **7**, VII, 259–284, Encyclopaedia Math. Sci., vol. 74. Springer, Berlin (1994)

78. Grauert, H., Remmert, R.: Theory of Stein Spaces. Classics in Mathematics. Springer, Berlin (2004)

79. Grauert, H., Riemenschneider, O.: Verschwindungssätze für analytische Kohomologiegruppen auf komplexen Räumen. Inventiones Math. **11**, 263–292 (1970)

80. Greene, R.E., Wu, H.: Embedding of open Riemannian manifolds by harmonic functions. Ann. Inst. Fourier (Grenoble) **25**(1), vii, 215–235 (1975)

81. Griffiths, P., Harris, J.: Principles of Algebraic Geometry. Wiley Classics Library. Wiley, New York (1994)

82. Hartogs, F.: Zur Theorie der analytischen Funktionen mehrerer unabhängiger Veränderlichen insbesondere über die Darstellung derselben durch Reihen, welche nach Potenzen einer Veränderlichen fortschreiten. Math. Ann. **62**. 1–88 (1906)

83. Hartogs, F.: Einige Folgerungen aus der Cauchyschen Integralformel bei Funktionen mehrerer Veränderlichen. Sitzungsberichte der Königlich Bayerischen Akademie der Wissenschaften zu München, Mathematisch-Physikalische Klasse **36**, 223–242 (1906)

84. Hartogs, F.: Über die aus den singulären Stellen einer analytischen Funktion mehrerer Veränderlichen bestehenden Gebilde. Acta Math. **32**, 57–79 (1909)

85. Hartshorne, R.: Cohomological dimension of algebraic varieties. Ann. Math. (2) **88**, 403–450 (1968)

86. Hartshorne, R.: Varieties of small codimension in projective space. Bull. Am. Math. Soc. **80**, 1017–1032 (1974)

87. Hartshorne, R.: Algebraic geometry. Graduate Texts in Mathematics, No. **52**. Springer, New York–Heidelberg (1977)

88. Harvey, F.R., Lawson, H.B.: Dirichlet Duality and the non linear Dirichlet problem on Riemannian manifolds. Commun. Pure Appl. Math. **62**, 396–443 (2009)

89. Harz, T., Shcherbina, N., Tomassini, G.: On defining functions and cores for unbounded domains. I. Math. Z. **286**(3–4), 987–1002 (2017)

90. Henkin, G.M., Leiterer, J.: Andreotti-Grauert theory by integral formulas. Progress in Mathematics, vol. 74, Birkhäuser, Boston (1988)

91. Hirzebruch, F.: Topological methods in algebraic geometry. Die Grundlehren der Mathematischen Wissenschaften, Band 131. Springer, New York (1966)

92. Ho, L.-H.: $\bar{\partial}$-problem on weakly q-convex domains complexes. Math. Ann. **290**, 3–18 (1991)

93. Horrocks, G., Mumford, D.: A rank 2 vector bundle on P^4 with 15,000 symmetries. Topology **12**, 63–81 (1973)

94. Hörmander, L.: L^2 estimates and existence theorems for the $\bar{\partial}$-operator. Acta Math. **113**, 89–152 (1965)

95. Hörmander, L.: An Introduction to Complex Analysis in Several Variables. D. Van Nostrand, Princeton, Toronto, London (1966)
96. Huckleberry, A.: Hyperbolicity of cycle spaces and automorphism groups of flag domains. Am. J. Math. **135**(2), 291–310 (2013)
97. Hunt, L.R. and Murray, J.J.: q-plurisubharmonic functions and a generalized Dirichlet problem. Michigan Math. J. **25**, 299–316 (1978)
98. Ioniţa, G.-I., Preda, O.: q-pseudoconvex and q-holomorphically convex domains. Math. Nachr. **292**(12), 2619–2623 (2019)
99. Joo, J.-C., Kim, K.-T., Schmalz, G.: On the generalization of Forelli's theorem. Math. Ann. **365**(3–4), 1187–1200 (2016)
100. Jöricke, B.: Envelopes of holomorphy and holomorphic discs. Invent. Math. **178**(1), 73–118 (2009)
101. Josefson, B.: On the equivalence between locally polar and globally polar sets for plurisubharmonic functions on $\mathbb{C}^n$. Ark. Mat. **16**(1), 109–115 (1978)
102. Kalka, M.: On a conjecture of Hunt and Murray concerning q-plurisubharmonic functions. Proc. Am. Math. Soc. **73**(1), 30–34 (1979)
103. Kiyosawa, T.: Some equivalent relations of q-convex domains in C^n. Sci. Rep. Tokyo Kyoiku Daigaku, Sec. A **10**, 245–249 (1969–1970)
104. Klimek, M.: Pluripotential Theory. London Mathematical Society Monographs. New Series, 6. Oxford Science Publications. The Clarendon Press, Oxford University Press, New York (1991)
105. Knorr, K., Schneider, M.: Relativexzeptionelle analytische Mengen. Math. Ann. **193**, 238–254 (1971)
106. Kodaira, K.: On a differential-geometric method in the theory of analytic stacks. Proc. Nat. Acad. Sci. U.S.A. **39**, 1268–1273 (1953)
107. Kodaira, K.: On Kähler varieties of restricted type (an intrinsic characterization of algebraic varieties). Ann. Math. (2) **60**, 28–48 (1954)
108. Kodaira, K., Spencer, D.C.: On deformations of complex analytic structures. I, II. Ann. Math. (2) **67**, 328–466 (1958)
109. Kohn, J.J.: Harmonic integrals on strongly pseudo-convex manifolds. I. Ann. Math. (2) **78**, 112–148 (1963)
110. Kohn, J.J.: Harmonic integrals on strongly pseudo-convex manifolds. II. Ann. Math. (2) **79**, 450–472 (1964)
111. Kohn, J.J., Nirenberg, L.: Non-coercive boundary value problems. Commun. Pure Appl. Math. **18**, 443–492 (1965)
112. Krantz, S.G.: Function Theory of Several Complex Variables, 2nd edn. 2001. American Mathematical Society, American Mathematical Society (1992)
113. Lelong, P.: Définition des fonctions plurisousharmoniques. C. R. Acad. Sci. Paris **215**, 398–400 (1942)
114. Lelong, P.: Les fonctions plurisousharmoniques. Ann. Sci. École Norm. Sup. (3) **62**, 301–338 (1945)
115. Levi, E.E.: Studii sui punti singolari essenziali delle funzioni analitiche di due o più variabili complesse. Annali di Mat. pura ed appl. (3) **17**, 61–87 (1910)
116. Levi, E.E.: Sulle ipersuperficie dello spazio a 4 dimensioni che possono cassere frontiera del campo di esistenza di una funzione analitica di due variabli complesse. Ann. Mat. Appl. (3) **18**, 69–79 (1911)
117. Matsumoto, K.: Pseudoconvex domains of general order and q-convex domains in the complex projective space. J. Math. Kyoto Univ. **33**(3), 685–695 (1993)
118. Matsumoto, K.: Boundary distance functions and q-convexity of pseudoconvex domains of general order in Kähler manifolds. J. Math. Soc. Jpn. **48**(1), 85–107 (1996)
119. Matsumura, S.: Asymptotic cohomology vanishing and a converse to the Andreotti-Grauert theorem on surfaces. Ann. Inst. Fourier (Grenoble) **63**, 2199–2221 (2013)
120. Miyazawa, K.: On the n-completeness of coverings of proper families of analytic spaces. Osaka J. Math. **39**(1), 59–88 (2002)

121. Morrey, C.B., Jr.: The analytic embedding of abstract real-analytic manifolds. Ann. Math. (2) **68**, 159–201 (1958)

122. Nakano, S.: Vanishing theorems for weakly 1-complete manifolds. Number theory, algebraic geometry and commutative algebra, in honor of Yasuo Akizuki. Kinokuniya, Tokyo, pp. 169–179 (1973)

123. Narasimhan, R.: The Levi Problem for complex spaces. Math. Ann **142**, 355–365 (1961)

124. Narasimhan, R.: Several Complex Variables. University of Chicago Press, Chicago (1971)

125. Nishino, T.: Sur les ensembles pseudoconcaves. J. Math. Kyoto Univ. **1**, 225–245 (1962)

126. Nishino, T.: Nouvelles recherches sur les fonctions entières de plusieurs variables complexes. II, Fonctions entières qui se réduisent à celles d'une variable. J. Math. Kyoto Univ. **9**, 221–274 (1969)

127. Nishino, T.: Function Theory in Several Complex Variables. American Mathematical Society, Providence (2001)

128. Norguet, F.: Sur les domaines d'holomorphie des fonctions uniformes de plusieurs variables complexes (Passage du local au global). Bull. Soc. Math. France **82**, 137–159 (1954)

129. Norguet, F., Siu, Y.- T.: Holomorphic convexity of spaces of analytic cycles. Bull. Soc. Math. France **105**(2), 191–223 (1977)

130. Ohsawa, T.: On complete Kähler domains with C^1-boundary. Publ. Res. Inst. Math. Sci. **16**(3), 929–940 (1980)

131. Ohsawa, T.: Completeness of noncompact analytic spaces. Publ. Res. Inst. Math. Sci. **20**(3), 683–692 (1984)

132. Ohsawa, T.: An interpolation theorem on cycle spaces for functions arising as integrals of $\bar{\partial}$-closed forms. Publ. Res. Inst. Math. Sci. **43**(4), 911–922 (2007)

133. Ohsawa, T.: Generalizations of theorems of Nishino and Hartogs by the L^2 method. Math. Res. Lett. **27**(6), 1865–1882 (2020)

134. Oka, K.: Note sur les familles de fonctions analytiques multiformes etc. J. Sci. Hiroshima Univ. Ser. A **4**, 93–98 (1934)

135. Oka, K.: Sur les fonctions analytiques de plusieurs variables VI. Domaines pseudoconvexes. Tôhoku Math. J. **49**, 15–52 (1942)

136. Oka, K.: Sur les fonctions analytiques de plusieurs variables. VII. Sur quelques notions arithmétiques. Bull. Soc. Math. France **78**, 1–27 (1950)

137. Oka, K.: Sur les fonctions analytiques de plusieurs variables IX. Domaines finis sans point critique intérieur. Jpn. J. Math. **23**, 97–155 (1953)

138. Oka, K.: A new method of generating pseudoconvex domains. Jpn. J. Math. **32**, 1–12 (1962)

139. Pawlaschyk, T.: On some classes of q-plurisubharmonic functions and q-pseudoconcave sets. Dissertation zur Erlangung eines Doktorgrades, Bergische Universität Wuppertal (2015)

140. Pawlaschyk, T., Shcherbina, N.: Foliations of continuous q-pseudoconvave graphs. to be published in Indiana Univ. Math. J. (2020). https://arxiv.org/pdf/2004.01797v1.pdf

141. Pawlaschyk, T., Zeron, E.S.: On convex hulls and pseudoconvex domains generated by q-plurisubharmonic functions, part I. J. Math. Anal. App. **408**, 394–408 (2013)

142. Pawlaschyk, T., Zeron, E.S.: On convex hulls and pseudoconvex domains generated by q-plurisubharmonic functions, part II. Bol. Soc. Mat. Mex. (3) **22**(2), 367–388 (2016)

143. Pawlaschyk, T., Zeron, E.S.: On convex hulls and pseudoconvex domains generated by q-plurisubharmonic functions, part III. Math. Anal. Appl. **471**(1–2), 73–87 (2019)

144. Peternell, M.: Algebraic and analytic cohomology of quasiprojective varieties. Math. Ann. **286**(1–3), 511–528 (1990)

145. Peternell, M.: Algebraische Varietäten und q-vollständige komplexe Räume. Math. Z. **200**(4), 547–581 (1989)

146. Poletsky, E.A.: Holomorphic currents. Indiana Univ. Math. J. **42**(1), 85–144 (1993)

147. Popa-Fischer, A.: A generalization to the q-convex case of a theorem of Fornæss and Narasimhan. Michigan Math. J. **50**(3), 483–492 (2002)

148. Remmert, R.: From Riemann surfaces to complex spaces. Matériaux pour l'histoire des mathématiques au XXe siècle (Nice, 1996), 203–241, Sémin. Congr., 3, Soc. Math. France, Paris (1998)

149. Richberg, R: Stetige streng pseudokonvexe Funktionen. Math. Ann. **175**, 257–286 (1968)
150. Rothstein, W.: Zur Theorie der analytischen Mannigfaltigkeiten im Raume von n komplexen Veränderlichen. Math. Ann. **129**, 96–138 (1955)
151. Schneider, M.: Über eine Vermutung von Hartshorne. Math. Ann. **201**, 221–229 (1973)
152. Schwarz, W.: Local q-completeness of complements of smooth CR-submanifolds. Math. Z. **210**, 529–553 (1992)
153. Shabat, B.V.: Introduction to complex analysis. Part II. Volume 110 of Translations of Mathematical Monographs. American Mathematical Society, Providence (1992)
154. Shcherbina, N.V.: On the polynomial hull of a graph. Indiana Univ. Math. J. **42**(2), 477–503 (1993)
155. Shcherbina, N.V.: Pluripolar graphs are holomorphic. Acta Math. **194**(2), 203–216 (2005)
156. Sibony, N.: Quelques problemes de prolongement de courants en analyse complexe. Duke Math. J. **52**, 157–197 (1985)
157. Skoda, H.: Prolongement des courants positifs fermes de masse finie. Invent. Math. **66**, 361–376 (1982)
158. Słodkowski, Z.: Analytic set-valued functions and spectra. Math. Ann. **256**, 363–386 (1981)
159. Słodkowski, Z.: The Bremermann-Dirichlet problem for q-plurisubharmonic functions. Ann. Scuola Norm. Sup. Pisa Cl. Sci. (4) **11**(2), 303–326 (1984)
160. Słodkowski, Z.: Local maximum property and q-plurisubharmonic functions in uniform algebras. J. Math. Anal. Appl. **115**(1), 105–130 (1986)
161. Stein, K.: Analytische Funktionen mehrerer komplexer Veränderlichen zu vorgegebenen Periodizitätsmoduln und das zweite Cousinsche Problem. Math. Ann. **123**, 201–222 (1951)
162. Suria, G.V.: q-pseudoconvex and q-complete domains. Comp. Math. **53**, 105–111 (1984)
163. Tadokoro, M.: Sur les ensembles pseudoconcaves généraux. J. Math. Soc. Jpn. **17**, 281–290 (1965)
164. Takeuchi, A.: Domaines pseudoconvexes infinis et la métrique riemannienne dans un espace projectif. J. Math. Soc. Jpn. **16**, 159–181 (1964)
165. Ueda, T.: Pseudoconvex domains over Grassmann manifolds. J. Math. Kyoto Univ. **20**(2), 391–394 (1980)
166. Vladimirov, V.S.: Methods of the Theory of Functions of Many Complex Variables. The MIT Press, Cambridge, London (1966)
167. Walsh, J.B.: Continuity of envelopes of plurisubharmonic functions. J. Math. Mech. **18**, 143–148 (1968/1969)
168. Yamaguchi, H.: Famille holomorphe de surfaces de Riemann ouvertes, qui est une variété de Stein. J. Math. Kyoto Univ. **16**(3), 497–530 (1976)
169. Yang, Xiaokui: A partial converse to the Andreotti-Grauert theorem. Compos. Math. **155**(1), 89–99 (2019)

Index

Printed in the United States
by Baker & Taylor Publisher Services

Printed in the United States
by Baker & Taylor Publisher Services